Careers, Entrepreneurship, and Diversity: Challenges and Opportunities in the Global Chemistry Enterprise

ACS SYMPOSIUM SERIES **1169**

Careers, Entrepreneurship, and Diversity: Challenges and Opportunities in the Global Chemistry Enterprise

H. N. Cheng, Editor
*Southern Regional Research Center
Agricultural Research Service
U.S. Department of Agriculture
New Orleans, Louisiana*

Sadiq Shah, Editor
*The University of Texas-Pan American
Edinburg, Texas*

Marinda Li Wu, Editor
*American Chemical Society
Washington, DC*

American Chemical Society, Washington, DC

Distributed in print by Oxford University Press

Library of Congress Cataloging-in-Publication Data

Careers, entrepreneurship, and diversity : challenges and opportunities in the global chemistry enterprise / H.N. Cheng, editor, Southern Regional Research Center, Agricultural Research Service, U.S. Department of Agriculture, New Orleans, Louisiana, Sadiq Shah, editor, The University of Texas-Pan American, Edinburg, Texas, Marinda Li Wu, editor, American Chemical Society, Washington, DC.
   pages cm. -- (ACS symposium series ; 1169)
   Includes bibliographical references and index.
   ISBN 978-0-8412-2970-9
   1. Chemistry--Vocational guidance--Congresses. 2. Chemistry--Vocational guidance--United States--Congresses. 3. Labor and globalization--Congresses. I. Cheng, H. N., editor. II. Shah, Sadiq, editor. III. Wu, Marinda Li, editor.
   QD39.5.C37 2014
   540.23--dc23
                          2014031161

The paper used in this publication meets the minimum requirements of American National Standard for Information Sciences—Permanence of Paper for Printed Library Materials, ANSI Z39.48n1984.

Copyright © 2014 American Chemical Society

Distributed in print by Oxford University Press

All Rights Reserved. Reprographic copying beyond that permitted by Sections 107 or 108 of the U.S. Copyright Act is allowed for internal use only, provided that a per-chapter fee of $40.25 plus $0.75 per page is paid to the Copyright Clearance Center, Inc., 222 Rosewood Drive, Danvers, MA 01923, USA. Republication or reproduction for sale of pages in this book is permitted only under license from ACS. Direct these and other permission requests to ACS Copyright Office, Publications Division, 1155 16th Street, N.W., Washington, DC 20036.

The citation of trade names and/or names of manufacturers in this publication is not to be construed as an endorsement or as approval by ACS of the commercial products or services referenced herein; nor should the mere reference herein to any drawing, specification, chemical process, or other data be regarded as a license or as a conveyance of any right or permission to the holder, reader, or any other person or corporation, to manufacture, reproduce, use, or sell any patented invention or copyrighted work that may in any way be related thereto. Registered names, trademarks, etc., used in this publication, even without specific indication thereof, are not to be considered unprotected by law.

PRINTED IN THE UNITED STATES OF AMERICA

Foreword

The ACS Symposium Series was first published in 1974 to provide a mechanism for publishing symposia quickly in book form. The purpose of the series is to publish timely, comprehensive books developed from the ACS sponsored symposia based on current scientific research. Occasionally, books are developed from symposia sponsored by other organizations when the topic is of keen interest to the chemistry audience.

Before agreeing to publish a book, the proposed table of contents is reviewed for appropriate and comprehensive coverage and for interest to the audience. Some papers may be excluded to better focus the book; others may be added to provide comprehensiveness. When appropriate, overview or introductory chapters are added. Drafts of chapters are peer-reviewed prior to final acceptance or rejection, and manuscripts are prepared in camera-ready format.

As a rule, only original research papers and original review papers are included in the volumes. Verbatim reproductions of previous published papers are not accepted.

ACS Books Department

Contents

Foreword xi

Preface xiii

1. Career Challenges and Opportunities in the Global Chemistry Enterprise 1
   Marinda Li Wu, H. N. Cheng, Sadiq Shah, and Robert Rich

Career Advancement Opportunities

2. Why Chemistry Still Matters 31
   John C. Lechleiter

3. Academic Opportunities and Considerations for a Successful Career in an Evolving Institution 39
   Jennifer S. Laurence

4. Industrial Careers: Obtaining a Job and Succeeding in Industry 49
   Joel I. Shulman

5. Career Opportunities in the Government Sector 61
   Kristin M. Omberg

6. From the Lab to a Career in Business Development – An Unexpected Career Transition 69
   Mark D. Frishberg

7. Connecting with a Most Powerful Ally of Science 83
   George Rodriguez

8. Alternative Chemistry Careers: Using Patent Assets for High Technology Commercial Innovation 93
   James L. Chao

9. And then the Wind Changed 107
   John Fraser

10. A Chemist Strategizing for Chemists: My Career in Associations and Societies 115
    Robert Rich

11. From Medical School Dropout to Editor-in-Chief of C&EN 129
    Rudy M. Baum

12. Consulting for Seniors: Having It Your Way 135
    Thomas R. Beattie

Innovation and Entrepreneurship

13. The American Chemical Society (ACS) Entrepreneurial Initiative 143
    Roger E. Brown, Elizabeth I. Fraser, Kenneth J. Polk, and David E. Harwell

14. Creativity, Innovation and the Entrepreneurship Nexus 149
    Sadiq Shah

15. Innovation and Entrepreneurship in the Chemical Enterprise 163
    Pat N. Confalone

16. Venture Capitalist Planning Is Irrelevant to Successful Entrepreneurship in Chemicals, Materials & Cleantech 173
    J. M. Ornstein and Hongcai "Joe" Zhou

17. A Tale of Academic Innovation to Job Creation in Rare Earth Extraction and Separation 183
    Neil J. Lawrence, Joseph Brewer, and Allen Kruse

18. Technical Entrepreneurship Serving Industry: A Personal Story 195
    Sharon V. Vercellotti and John R. Vercellotti

Diversity and Inclusion

19. A Top-Down Approach for Diversity and Inclusion in Chemistry Departments 207
    Rigoberto Hernandez and Shannon Watt

20. Diversity and Departmental Rankings in Chemistry 225
    Cedric Herring

21. Promoting Diversity and Inclusivity at Rose-Hulman Institute of Technology 237
    Luanne Tilstra

22. Improving Transparency and Equity in Scholarly Recognition by Scientific Societies 245
    Erin L. Cadwalader and Amanda C. Bryant-Friedrich

23. American Indian Science and Engineering Society (AISES): Building a Successful Model for Diversity and Inclusion 255
    Mary Jo Ondrechen

24. Career Opportunity Challenges for Asian American Chemistry
    Professionals and Scientists 265
    Guang Cao

25. Diversity and Inclusion from the Global Perspective 273
    Teri Quinn Gray

Editors' Biographies 283

Indexes

Author Index 287

Subject Index 289

Foreword

This is the second of three ACS Symposium books stemming from my initiatives in the ACS presidential succession. The first book was based on my presidential symposia at the 2013 national ACS spring meeting in New Orleans, Louisiana when I was ACS President. It was published online in March 2014 (pubs.acs.org/isbn/9780841229389) and is now available in hard copy as well. Its 22 chapters share perspectives of ten presidents representing chemical societies from Europe, Asia, Africa, and the Americas as well as thought leaders from the USA representing academia, industry, government, and small business.

This second ACS Symposium book is based on three presidential symposia at the 2013 national ACS fall meeting in Indianapolis, Indiana. It covers the experiences and perspectives of speakers at three symposia focused on Careers, Entrepreneurship, and Diversity in the global context. All three symposia were important focus areas of my presidential initiatives.

The first symposium topic on *"Career Advancement Opportunities"* offers ideas and career advice for new graduates as well as seasoned mid-career chemistry professionals interested in exploring new career options. Many of the authors started with an interest in science or as bench chemists in the laboratory, but pursued different routes to rewarding careers. Career stories range from becoming the successful CEO of Eli Lily Company (Dr. John Lechleiter, Chapter 2) to becoming Editor-in-Chief of the *Chemical & Engineering News* magazine (Rudy Baum, Chapter 11). The second symposium topic, *"Innovation and Entrepreneurship,"* provides examples of successful entrepreneurs who started as chemists. By sharing their career experiences, we hope this book will catalyze your own ideas on how to bring innovation to the market as a chemist.

The ACS Committee on Minority Affairs (CMA), chaired by Al Ribes of Dow Chemical, organized the third symposium topic, *"Impact of Diversity and Inclusion."* I have always strongly supported diversity and inclusivity — long before I was elected ACS President — so it naturally became one of my presidential initiatives. Indeed, I was recently invited to talk about the role of *diversity in science* for the Council of Scientific Society Presidents (CSSP) annual meeting in May 2014 in Washington D.C. My presentation, which I called *"Partners for Progress and Prosperity: Celebrating Diversity Together!"* is one I give both here in the USA as well as overseas to raise awareness of the importance of diversity and how it enriches our discussions.

The importance of raising awareness and appreciation for both diversity and globalization cannot be underestimated if we are to thrive in the global chemistry (and scientific) enterprise of the 21st century. A better understanding of many diverse cultures and backgrounds is essential for effective teamwork and *"Partnering for Progress and Prosperity!"* More details are provided in my presidential article in *C&EN*, Jan. 7, 2013, pg. 2.

Many thanks go to the many ACS colleagues, ACS staff, and members of the global chemistry enterprise who provided input and help throughout my years in the ACS Presidential Succession. *Extra special thanks* go to my co-editors and Presidential Task Force co-chairs — Dr. H.N. Cheng and Dr. Sadiq Shah — for partnering with me on my ACS Presidential initiatives and these books! I am also grateful for the enduring support from my family — my dear mother, Tsun Hwei Li, now 94 years old, plus my two fabulous children who are successful young adults with their own satisfying careers (my talented son Will who designed my presidential logo for "Partners for Progress and Prosperity" on the front cover of this book and my amazing daughter, Lori, who married her wonderful Stanford classmate, Evan), and most of all to my fantastic partner in life, my husband Norm Wu!

Thanks to all as we continue to "Partner for Progress and Prosperity!"

Marinda Li Wu
2013 ACS President
July 2014

Preface

This book is the result of three Presidential Symposia held at the 246th National Meeting of the American Chemical Society (ACS) in Indianapolis in September 2013. These symposia focused on: 1) Career Advancement Opportunities, 2) Innovation and Entrepreneurship, and 3) Impact of Diversity and Inclusion. The speakers at the symposia were all accomplished leaders of the global chemistry enterprise, and their talks were very well received. Subsequently, the speakers were invited to contribute chapters to this book. Most of them did, and this book is the result of their hard work.

A total of 25 chapters are compiled in this book. Consistent with the symposia themes, this book is divided into three sections: 1) Career Advancement Opportunities, 2) Innovation and Entrepreneurship, and 3) Diversity and Inclusion. Chapter 1 is an overview chapter that summarizes the contents of the various chapters and also provides additional information on the different career paths for chemistry professionals. This first chapter also outlines the latest resources available at ACS for job search and career development and management.

This book is intended to be useful for all scientists and engineers in general, and particularly chemists, biochemists, chemical engineers, biochemical engineers, and others in chemistry-related fields. People interested in exploring various jobs and careers may find Chapters 1–12 especially useful. Those scientists and engineers interested in starting their own businesses may find useful insights in Chapters 13–18. Readers interested in learning more about diversity and inclusion in the scientific workplace may find helpful information and case histories in Chapters 19–25. For anyone interested in enriching their knowledge about the chemical profession and various career paths and opportunities, the perspectives and personal experiences shared by contributors of many of these chapters should be of great interest. Indeed the global chemistry enterprise is changing, and it is important for chemistry professionals and students alike to be aware of the challenges and the opportunities available today in order to manage their careers accordingly.

We appreciate the efforts of the authors who took time to prepare their manuscripts and our many reviewers for their time and effort during the peer review process. We thank many ACS staff members, who ably assisted in the various tasks related to this book, particularly Robert Rich, Steven Meyers, Elizabeth McGaha, Gareth Edwards, David Harwell, Joy Titus-Young, Max Saffell, and Bradley Miller. Thanks are also due to the personnel at ACS Books, especially Tim Marney, Arlene Furman, and Bob Hauserman. We hope that you will find this book a valuable resource in helping to successfully navigate your own career path in the Global Chemistry Enterprise.

On the cover: The photograph in the upper left is from Oregon Department of Transportation, Diversity Conference 2013. https://www.flickr.com/photos/oregondot/10314223086/. The photograph in the lower right is from U.S. Army RDECOM, Army Scientists energize battery research. https://www.flickr.com/photos/rdecom/7336830168/in/set-72157629845817908/. Both are covered under the Creative Commons license http://creativecommons.org/licenses/by/4.0/.

H. N. Cheng
Southern Regional Research Center
USDA – Agricultural Research Service
1100 Robert E. Lee Blvd.
New Orleans, LA 70124, USA

Sadiq Shah
The University of Texas-Pan American
1201 W. University Drive,
Edinburg, TX 78539, USA

Marinda Li Wu
American Chemical Society
1155 Sixteenth Street, N.W.
Washington, DC 20036, USA

Chapter 1

Career Challenges and Opportunities in the Global Chemistry Enterprise

Marinda Li Wu,[1,*] H. N. Cheng,[2,*] Sadiq Shah,[3,*] and Robert Rich[4,*]

[1]2013 ACS President, American Chemical Society, 1155 Sixteenth St., N.W., Washington, DC 20036
[2]USDA Agricultural Research Service, Southern Regional Research Center, 1100 Robert E. Lee Blvd., New Orleans, Louisiana 70124
[3]The University of Texas Pan American, 1201 W. University Drive, Edinburg, Texas 78539-2999
[4]American Chemical Society, 1155 Sixteenth St., N.W., Washington, DC 20036
*E-mails: marindawu@gmail.com; hn.cheng@ars.usda.gov; sadiq@utpa.edu; r_rich@acs.org

> This chapter serves as an overview of the various career challenges and opportunities faced by chemistry professionals in the 21st century in the global chemistry enterprise. One goal is to highlight a broad spectrum of career paths, including non-traditional careers, and to showcase examples of chemists who have successfully followed these paths. Successful examples of entrepreneurship by chemistry professionals are also shared as one of many different career advancement opportunities to explore. All of these career paths use chemistry or a scientific background and analytical problem-solving skills to organize information, identify creative solutions, and make meaningful contributions to society. Another goal is to cover topics on diversity in the workplace as it relates to these paths. Included in this article are global employment trends and services that the American Chemical Society offers to help chemists to search for jobs and to manage their careers.

© 2014 American Chemical Society

Introduction

The global chemistry enterprise is robust but constantly changing. Traditionally, chemistry professionals are educated to work in industry, academia, or government laboratories. A small (and growing) number of people opt to pursue an entrepreneurial path to launch their businesses, which may include analytical testing, product development, contract manufacturing, or consulting. There are also many people who are trained in chemistry but who pursue a non-laboratory career, such as business or nonprofit management, sales and marketing, patent law, regulatory and environmental compliance, and science writing. All of these career paths use the chemistry or scientific background and the analytical problem solving skills to organize information, identify creative solutions, and make meaningful contributions to society.

Another factor that affects jobs and careers in recent years is the increasing globalization of the chemistry enterprise. This trend has a noticeable impact on employment and career development of chemists and chemical engineers in the U.S. and worldwide. To some people it may be advantageous; to others it may be a disadvantage. Much has been written about this trend (*1–3*).

In response to the changes in the chemistry enterprise, ACS President-Elect Marinda Li Wu appointed a Presidential Task Force in 2012 to study both the challenges and opportunities related to jobs, advocacy and globalization. After considerable analysis and discussion, the Task Force developed seven recommendations (*4*), as shown below. More background and details on the development of these recommendations by the ACS Presidential Task Force "Vision 2025" can be found in the first ACS Symposium book based on President Wu's presidential symposia called "Vision 2025: How to Succeed in the Global Chemistry Enterprise" (*4*).

1. Better educate ACS members about the critical elements necessary for success in a broad spectrum of career paths.
2. Strengthen ACS efforts to support entrepreneurship.
3. Engage and equip members with enhanced advocacy tools and training so that they can proactively contact their legislators to improve the business climate and aid jobs creation.
4. Explore with U.S. and global stakeholders the supply and demand of chemists/jobs to bring them to a better equilibrium.
5. Collaborate with others, including chemical societies around the world regarding public communication, education, advocacy, chemical employment, and other topics of mutual interest.
6. Provide information, resources, advice, and assistance to ACS members interested in global job opportunities.
7. Expand ACS support for chemists and chemistry communities worldwide.

As a follow-up to the Task Force recommendations, three Presidential symposia were organized at the ACS national meeting in Indianapolis in September 2013 on 1) Career Advancement Opportunities, 2) Innovation and

Entrepreneurship, and 3) Diversity and Inclusion. The speakers in these three symposia provided excellent perspectives on career and workplace issues. Subsequently most of them contributed chapters to this symposium book (5–28).

This article is designed to serve as an overview of the career challenges and opportunities as well as a summary of diversity and inclusion activities in the chemistry enterprise. It also highlights the significant contributions made by the various authors in this book. Some pertinent global employment trends are described, and services that the American Chemical Society offers to help chemists to search for jobs and to manage their careers are also included.

To facilitate the discussion of chemistry careers, some of the 2013 employment data on ACS members are displayed in Table 1. Note that about 54% of ACS members work in industry and 37% in academia. About 30% of ACS members are female. The racial and the ethnic make-up among ACS members is also indicated in the table.

Table 1. % Breakdown of ACS members by categories. Data taken from 2013 ACS Employment and Salary Survey (29)

Degree Level	Bachelor's	Master's	PhD	Total
By Gender				
Men	63.6	64.2	73.3	70.2
Women	36.4	35.8	26.7	29.8
By Employer				
Industry	82.9	68.4	44.1	54.3
Government	7.7	7.6	7.6	7.6
Academia	8.2	21.8	46.8	36.5
Self-employed	1.2	2.1	1.5	1.6
By Race				
American Indian	0.3	0.3	0.2	0.2
Asian	4.0	8.4	12.0	10.1
Black	3.5	2.2	2.3	2.5
White	89.6	86.8	83.8	85.3
Other	2.7	2.3	1.7	1.9
By Ethnicity				
Hispanic	4.2	3.3	3.7	3.7

Career Challenges

For a job seeker in the United States, 2009-2013 were difficult years (Table 2). The U.S. economy was in recession or barely growing. Many large chemical companies in the U.S. tended to shy away from fundamental or longer term research. Traditional research & development (R&D) chemistry jobs in large U.S. companies became less prevalent. Hiring became more selective and restricted to specific needs. Many pharmaceutical companies merged or downsized. In the mean time, the U.S. government faced increasing competition for its financial resources and debated the merits of R&D spending. Support for academic R&D was constrained, with negative impact on academic jobs.

Table 2. Chemistry employment data (in %), 2009-2013 (*30*)

Workforce employment	2009	2010	2011	2012	2013
Full-time	90.3	88.3	89.8	90.0	91.3
Part-time	3.2	3.9	3.8	3.2	2.7
Postdoc	2.5	4.0	1.8	2.6	2.5
Seeking employment	3.9	3.8	4.6	4.2	3.5

The unemployment data in chemistry-related jobs in 2009-2013 are given in Table 3. BA/BS unemployment reached an all-time record of 6.4% in 2011. Fortunately, the trend has improved in 2013.

Table 3. Unemployment for all degree levels in chemistry-related job (in %), 2009-2103 (*30*)

Unemployment	2009	2010	2011	2012	2013
BA/BS	5.6	5.1	6.4	5.9	4.6
MS	4.2	4.8	5.2	5.4	4.7
PhD	3.3	3.2	3.9	3.4	3.0

Employment skill sets in the traditional sense (e.g., book knowledge, lab skills, familiarity with instrumentation, and line management) are no longer sufficient for many of the jobs available today. Thus, many new employees need to acquire stronger "soft" skills, such as communication, salesmanship, negotiation, and grantsmanship (*31*). There is an increasing need to be able to work effectively as a part of diverse teams. Furthermore, chemistry plays an important role in an increasingly multidisciplinary global enterprise. Thus, cutting edge and significant frontier science often takes place at the interfaces of different disciplines. Visit http://strategy.acs.org to note that one of the Core Values of ACS is Diversity and Inclusion, highlighting the "advancement of chemistry as a global multidisciplinary science."

Yet another factor that has affected jobs and careers in recent years is the accelerating globalization of the chemistry enterprise. From the U.S. perspectives, for many years international graduate and undergraduate students in science and engineering have come to the U.S. to study; this trend continues although other countries are attracting increasingly larger number of international students (*32*). Some of them may stay and help strengthen the nation's chemistry enterprise. Faculty members have the option of collaborating with international colleagues. They can access instruments and facilities worldwide, and many professors can go on sabbaticals to enrich their careers. Likewise, industry can benefit from decreased cost of manufacturing overseas, easier access to international markets, and access to global talent.

From the U.S. perspective, a disadvantage of globalization is its effect on jobs and career development in the U.S. As jobs are outsourced overseas, domestic jobs tend to decrease. At the same time foreign-born scientists who stay in this country may compete with domestic students for the decreasing number of jobs. One way to mitigate this situation is for U.S. scientists to explore opportunities for jobs overseas. Another way is to encourage entrepreneurship in the U.S. in order to take advantage of technological innovation domestically. It may also be beneficial to approach policy makers and opinion leaders to advocate for more support for Science, Technology, Engineering, and Mathematics (STEM) research and education in the U.S.

Of course, it is not sustainable to produce more chemistry graduates when chemical jobs are diminishing. Another ACS Task Force was initiated by President Wu on supply/demand of chemists and jobs in the United States. Members representing stakeholder ACS committees with the Committee on Economic and Professional Affairs (CEPA) taking the lead were appointed during the summer of 2013. A Task Force report of its findings is expected by the fall of 2014.

In view of these developments, a chemistry professional needs to be flexible in order to adapt to the ever-changing global chemistry enterprise The idea of working an entire career in one country, or with one employer, has become increasingly obsolete and limiting. It is useful to consider the full spectrum of job opportunities as one's experience base expands. Some different career opportunities to consider exploring are shared in the next section.

Career Opportunities

Academic Teaching Career

First, the opportunities in an academic teaching career may be addressed. There are several options. These include job opportunities in primary and secondary schools, community colleges, comprehensive universities, research intensive universities, and private four year colleges. Teaching responsibilities are a common theme among all. There are distinct differences depending upon the type of institution, however. A table showing the breakdown of ACS members working in different academic institutions is shown in Table 4.

Table 4. % ACS academic members working in different academic institutions (30)

Institution type (highest degree awarded)	2009	2010	2011	2012	2013
Associate's	7.8	7.8	7.9	7.5	6.5
Bachelor's	20.8	22.8	23.9	24.8	24.8
Master's	7.8	7.1	6.7	7.7	8.0
Doctorate	45.1	45.2	41.7	40.9	43.8
University, medical, or professional school	9.4	6.3	8.5	8.0	8.5
High school	7.1	8.9	8.0	7.8	6.1
Other academic	2.1	1.9	3.3	3.3	2.4

One of the job opportunities for chemists and chemical engineers is to teach at primary and secondary schools (K-12 levels). There seems to be a continuing demand for teachers, particularly in the STEM areas (33). Thus, this is an excellent profession for someone who likes to work with children and who has an interest in teaching.

Faculty members at community colleges, for the most part, have heavy teaching loads and essentially no research expectations. If a faculty member is interested in research, collaboration with a nearby research institution becomes a viable option. It should be pointed out that students at the community college are typically not engaged in research; however, the National Science Foundation and the National Institutes of Health now fund programs to enable community college faculty members to engage their students in collaborative research with faculty at larger universities (34, 35). These programs are designed to support the movement of community college students to four year institutions to finish an undergraduate degree and potentially move on to seek graduate degrees. The research activities are designed to inspire and motivate students to advance their education by finishing four year degrees and graduate degrees.

The goal of these NSF/NIH programs is to support a steady pipeline of skilled workforce for STEM fields (36). Typically, the tenure decision at community colleges is based on teaching responsibilities. Community colleges also hire a number of adjunct faculty on an annual contract or on semester basis, with no long-term employment commitment by the institution. This is on an as-needed basis. The majority of the adjunct faculty are working chemists and they teach courses at community colleges beyond their respective full time jobs, or these may be retired chemists or others who used to work in industry.

Private four year institutions do not typically have extensive research requirements, though this is changing as the value of undergraduate research experiences to scientific education is recognized. There is often a heavy teaching workload for faculty. A growing number of private institutions encourage some undergraduate research for all science majors. They may seek gifts from external sources to financially support such initiatives. Sometimes faculty members submit proposals for external funding to federal and private agencies (such as the ACS Petroleum Research Fund or the Camille and Henry Dreyfus Foundation) to support such research activity on campus. The faculty members can have a working research relationship with faculty at a comprehensive or a research university within the region to access sophisticated research instrumentation for collaborative research efforts, though many leading private colleges now have such equipment on their own campus. Collaborations also allow their students to be introduced to possible career paths and the opportunities to pursue graduate degrees at the comprehensive or research intensive universities.

Comprehensive and/or research intensive universities typically require faculty members to maintain an active research program. Thus, in addition to teaching responsibilities (although with a reduced teaching load), faculty members are expected to secure external research funding to support their research agenda. Typically faculty members are hired and provided some start-up funding to establish their laboratories and launch their research endeavors. Depending upon the institution and level of research intensity, start-up funding can vary from $20K to millions. In return for this investment, the universities expect faculty to then successfully secure external research funding. This is an important factor in whether they will be granted tenure after 5-7 years at the institution.

In this book, the chapter by Laurence (6) provides an excellent example of a successful academic career. It also contains many helpful tips on how to thrive at a research university.

Academic Administrative Career

Some faculty may choose to pursue an administrative path within a university. This is typically possible after a few years of experience at the university and after tenure is granted. Faculty may move into the position of Chemistry Department Chair. Then, after successful experience as chair of a department, they may move into administrative roles in the Office of the Dean of the College of Science as an Assistant or Associate Dean and in some cases directly into the Dean of the College position or perhaps at another institution.

The college Dean serves as the chief academic officer for the college. Some of the deans may eventually move into a more senior administrative role in the central academic affairs office as an Assistant or Associate Vice President of Academic Affairs, then Assistant or Associate Vice Provost and then eventually Provost and Vice President for Academic Affairs, serving as the Chief Academic Officer for the University, and finally becoming the President of the university in some cases (37).

However, generally they move into the Provost or the President level position at a different university. It is relatively rare that a faculty member moves into such a senior level position within the same university. Senior administrative positions are typically filled externally to bring new ideas to the institution. The Vice

President or the Vice Provost serves as the chief research officer at the institution. This role requires a broad experience base and understanding of seeking external funding for the institution, awards administration, compliance with federal regulations related to grant administration, research involving human subjects, animal care and use, biosafety issues, export control issues, intellectual property protection and commercialization experience, addressing and managing conflicts of interest issues, business contracts to interface with industry for collaborative research through agreements. In such positions, one is involved in interfacing with faculty from all colleges and all disciplines as well as other divisions on campus, and must enjoy working with faculty, staff and external communities.

The Office of Research is a central office for the entire research university to offer services and address issues related to all aspects of research. Another administrative path is becoming the Dean of Graduate School with responsibilities to manage the graduate student recruitment, establishing degree programs in collaboration with all colleges at the university. This serves as a central point for graduate programs at the university and deals with all aspects of the Graduate School, including accreditation of degree programs. There are a number of administrative job opportunities in academia; however, it is for those who enjoy working with people conceiving new programs, implementing them, and monitoring the success of these student-centered programs.

Some faculty with a very active research agenda and extensive external funding may become directors of research centers or institutes on campus. Their role is then to manage and grow the research activities at the center by securing external funding and managing the financial and technical well being of the center or the institute. This experience can also prepare faculty for advancing their careers in central administration later. Such positions offer an experience related to the research accomplishments to make a broader impact and recognition for the institution. Responsibilities may include managing the finances like a business, dealing with people, developing a vision for the center, conceiving new initiatives, building partnerships and consensus, and advocating for and promoting the center and its role.

Other laboratory career opportunities include research labs affiliated with federal or state government or free-standing research institutes. Free-standing research institutes may be independent non-profit organizations or affiliated with a university (*38*).

Industrial Career

Chemists and chemical engineers find employment in different industries. A list of employer segments that hire ACS members is given in Table 5.

Table 5. Type of industrial employers amongst ACS industrial members (30)

2013 Type of Employer (Industry Only)	%
Aerospace/auto/transportation	2.5
Agricultural chemicals	2.3
Basic commodity chemicals	2.9
Biochemical products	2.2
Building materials	0.9
Coatings/paints/inks	4.1
Electronics/computers/semiconductors	2.7
Food	3.6
Instruments	3.1
Medical devices/diagnostic products	3.6
Metals/minerals	1.1
Paper	0.5
Personal care	1.7
Petroleum/natural gas	4.7
Pharmaceutical products	22.2
Plastics	2.9
Rubber	0.7
Soaps/detergents/surfactants	1.3
Specialty/fine chemicals	8.3
Textiles	0.3
Other manufacturing	5.8
Analytical service/testing laboratory	3.1
Biotech research firm	4.1
Independent or contract research firm	2.6
Hospital or clinical laboratory	0.4
Non-profit organization	1.9
Private utility company	0.2
Professional services - scientific/engineering/law	5.0
Research institution	2.8

Continued on next page.

Table 5. (Continued). Type of industrial employers amongst ACS industrial members (*30*)

2013 Type of Employer (Industry Only)	%
Scientific temporary or personnel agency	0.1
Other non-manufacturing	2.5

In corporate settings, chemists play an important role in supporting the research agenda and sustaining existing product lines. There are multiple career paths such as research and development (R&D) laboratories that have distinct sections or departments dedicated to a specific aspect of the business. A few companies have central research laboratories where basic and applied research is done and analytical laboratories for testing and analysis. Industrial chemists also work in quality control laboratories to support manufacturing, and process development laboratories to address the scale-up of new products from the bench scale to millions of gallons or pounds for commercial production. Other chemists and chemical engineers may get involved with commercial development, marketing, sales, manufacturing, safety and regulatory affairs, patents/licensing, and new business development. A breakdown of the different work functions that ACS industrial members are engaged in is summarized in Table 6.

A majority of new graduates start their careers in central research laboratories, R&D laboratories, process development, or technical support. Depending upon the business focus of the company, the R&D laboratories may be composed of synthetic chemistry laboratories, analytical services, application development, product testing, engineering and process development, and (for pharmaceutical companies) preclinical or clinical studies. Technical services and commercial development involving frequent interactions with marketing, sales and customers offer additional career routes.

Some chemistry professionals may be involved in addressing regulatory issues related to the products (e.g., safety and toxicity, disposal of byproducts, possible contaminants into air and water, and special handling during manufacturing), thus dealing not directly with laboratory work, but with regulatory issues linked with regulatory bodies such as FDA, EPA, and other agencies. They typically advise company employees on regulatory compliance related to the products and also facilitate filing applications with regulatory bodies for the various approval processes of new or improved products.

Generally Ph.D.-level scientists may spend some time in the R&D laboratories and then after some experience may move into management, technical service, process development, or quality control. In large corporations, there are often opportunities to move around in the company from R&D to marketing, sales, manufacturing, or business development. However, it is sometimes difficult to move back into a research role from non-laboratory functions because of technological advancements.

Table 6. % Breakdown of industrial work functions in chemical industry according to ACS member survey (30)

2013 Work Function (Industry Only)	%
Analytical Services	11.0
Chemistry information services	1.1
Computer programming, analysis, design	1.2
Consulting	2.7
Forensic analysis	0.7
General management (non-R&D)	4.0
Health and safety/regulatory affairs	3.2
Marketing, sales,	5.2
Patents, licensing, trademarks	2.0
Production, quality control	6.9
R&D: Applied research	38.5
R&D: Basic research	8.6
R&D: Management	10.9
Teaching or training	0.6
Other	3.4

The business development role is most suited for those who have gained significant experience with the company's product lines and have demonstrated knowledge and interest on the business side of the company. More importantly, they need to have the skills (and the personality) to interact with senior management and the marketing and sales force, but they are also able to speak the language and priorities of the research arm of the company.

Most companies offer two career ladders. One is the research ladder within research & development, where one can progress from a chemist (scientist) to senior chemist (scientist) to a research fellow, to senior research fellow. Frequently the Vice President of R&D serves as the Chief Scientific Officer of the company.

On the management side within R&D, one can start with the project manager role and progress to group leader, manager and director of a laboratory or research section with a reporting line to the Vice President of R&D. People in the management track need a broader understanding of the business of the company, management aptitude, communication and people skills. They need to interact with senior management related to technical issues in a business context. Thus an interest on the business side is a prerequisite. These roles require working with the business side, such as with marketing, with business development, and sometimes with sales and customers. As needed, they may also interact with manufacturing, regulatory affairs and finance.

Readers interested in more information on an industrial career may want to read Chapters 2, 4, and 6 of this book. In their chapters, Lechleiter (5) and Shulman (7) point out the key success factors in an industrial career. In addition to technical and problem solving skills, a successful industrial scientist needs to have soft skills and aptitudes involving teamwork, communication, leadership, and willingness to take risks. In his chapter, Frishberg (9) gives a good example of a career transition from research to business development and the skills needed to do this successfully.

Small Businesses and Entrepreneurship

A subset of an industrial job market comprises the small chemical businesses. Many small businesses may not be well known to the general public, but they play important roles as employers and as breeding grounds for innovation. The chapters in this book by Confalone (18) and Shah (17) point out the importance of entrepreneurship, creativity, and innovation to economic growth. According to BLS data, over 50% of chemistry graduates are hired by small businesses (defined as companies with less than 500 employees) (39).

Entrepreneurship can start at any age and at any stage of a career. Some fresh graduates may choose to pursue an entrepreneurial path for their careers by launching their own businesses. Mike Strem of Strem Chemicals is a good example, who started his business fresh out of graduate school (40). In some cases professors may launch start-up companies, based on the research that they may have conducted and then licensed the intellectual property (IP) from the university. An example is given in this book by Ornstein and Zhou (19). They saw an opportunity in metal-organic framework molecules and started a new company. Sometimes a person may start a business at mid-career; a good example is W. L. Gore, whose company is now well known for Gore-Tex® and many other products (41).

There are many other examples of successful entrepreneurs in this book. For many years Vercellotti and Vercellotti (21) have been running a carbohydrate business in Louisiana. Lawrence, Brewer, and Kruse (20) have co-founded a company doing rare earth extraction and separation. Beattie (15) and Rodriguez (10) are in the consulting business. Beattie's chapter (15) provides a good account of the challenges and the opportunities of consulting for seniors.

In a start-up company or a small business, a partner (or owner) often wears multiple hats and may be involved in several roles blended together. One must be flexible to adapt to different needs. The reward is that one will gain various experiences and learn about the different facets of the business and therefore become much more attuned to future opportunities. A chemist working for a small business will also gain broad experiences, which will better prepare him or her for future job opportunities. This is especially useful if the employee is interested in starting his or her own business in the future.

It may be noted that in some high tech areas (e.g., biotech and pharma), the risks and the rewards can both be high. Thus, the successful development and clinical trial of a drug molecule can bring about huge rewards, but the R&D investments can be substantial before the drug reaches the clinical stage. There is always an effort on the parallel path to seek additional investment in the company to fund the research efforts to get to the desired exit stage for the company. It is often a race to develop product lines and market them before the money (initially from investors) runs out.

The ACS is keenly aware of the need for entrepreneurship in order to stimulate innovation in the chemistry enterprise (*16*, *18*, *42*). It has started a pilot Entrepreneurial Initiative in 2011-2013, which has been well received. An ACS Entrepreneurial Resource Center is now available to help chemistry entrepreneurs. More information is given in the chapter by Brown, et al (*16*).

Career in Government

According to ACS employment surveys (Table 1), less than 8% of ACS members work for government agencies at federal, state and local levels. In general, the compensation for the government workers tends to be somewhat lower than the compensation in the industrial sector but higher than that in the academic sector (*29*). Many government agencies often use contractors, postdocs, and/or temporary workers to do the actual work. Salaries and benefits can vary depending on the contracting firms, the terms of contracts, and the nature of work.

An advantage of government jobs is that the information on most job openings is posted on the web or available upon request. Thus, all U.S. federal government job openings are posted on www.usajobs.gov. A prospective job seeker can enter specific job parameters (degree level, discipline, grade/salary level, citizenship requirement) and search for available job openings. State and local governments have different websites and can be searched separately. This does not, however, diminish the importance of networking to explore posted opportunities, since some of the posted jobs are not real openings, perhaps being held for a specific individual or requiring a particular status or experience.

A wide range of jobs are available in the government sector. Some government labs are well known, such as National Institutes of Health (NIH) and National Institute of Standards and Technology (NIST). A list of 39 Federally Funded Research and Development Centers (FFRDCs) is available on the web (*43*). Examples are Argonne, Brookhaven, Oak Ridge, Los Alamos, Sandia, Jet Propulsion Laboratory, National Renewable Energy Laboratory, Lawrence Berkeley and Lawrence Livermore.

In addition, numerous government agencies operate their own research laboratories. Examples are Department of Defense (DOD), National Aeronautics and Space Administration (NASA), Department of Homeland Security (DHS), Department of Agriculture (USDA), Environmental Protection Agency (EPA), Food and Drug Administration (FDA), and intelligence agencies such as the Federal Bureau of Investigation (FBI) and the Central Intelligence Agency (CIA).

There are also many non-laboratory government jobs for science majors. A good example is the U.S. Patent and Trademarks Office (USPTO), which has hired a large number of patent examiners in recent years (*44*). Other non-laboratory jobs include policy analysis, regulatory compliance, foreign affairs, program management, human resources, and staff support. An excellent summary of the opportunities and the nature of jobs in the government sectors has been provided by Omberg in her chapter (*8*).

Non-Traditional Careers

In addition to the traditional job categories of academia, industry, and government, there are many non-traditional career options open to chemistry professionals. Some examples are shown below.

Forensic Science

This term refers to the scientific methods of gathering and examining information about the past, particularly in reference to law enforcement. A chemistry degree holder is particularly suited to be a crime laboratory analyst. With more training, a chemist can also be a crime scene examiner, forensic engineer, or forensic archeologist.

Health, Environment, and Safety

In academia, industry, and government research facilities, often one or more people are needed to ensure laboratory safety, proper use and documentation of materials (including chemicals and biological agents), safe disposal of materials and equipment, compliance with government regulations, and employee health issues. This is a viable job option for chemistry degree-holders. While this job function does not require someone to work full-time on the laboratory bench, a good knowledge of laboratory procedures and chemical safety is a distinct advantage.

Technology Transfer

The Office of Technology Transfer at a university typically licenses the technologies developed within the university to anyone interested. Sometimes an outside company or an entrepreneur may show interest; sometimes the professors (and/or students) who develop the technology may want to commercialize the technology themselves. The IP is typically patented by the university and licensed to the entrepreneur(s), with some due diligence by the Technology Transfer Office. Federal research laboratories also have Offices of Technology Transfer, and IP can be licensed from these laboratories to launch businesses. Sometimes large corporations may license technology to outgoing employees when there is a downsizing or re-organization underway. However, this depends on the interest of the outgoing employees and their ability to commercialize the IP. Sometimes, chemists upon exit from industry have enough experience to launch their own contract R&D firms or analytical services businesses.

University Technology Transfer Offices are often looking for individuals who have the appropriate technical background and have gained business experience in industry. To staff an Office of Technology Transfer requires experience with both research and business. Thus, those who have been working in business development in corporations have just the right backgrounds for these positions, and the salaries are very competitive with industry.

To learn more about these opportunities, the reader is referred to the Association of the University Technology Transfer Managers (AUTM) website (45). AUTM is the professional organization of technology transfer officers. They typically have job fairs at their national meetings, where potential employers can meet with candidates for initial interviews. This may lead to an onsite interview. It may also be an opportunity to talk to professionals from various organizations about their open positions. It is possible for students to take internship positions in Technology Transfer Offices to learn and then move into a full time position. Readers interested in this career path may want to read the chapter by Fraser (12), a former president of AUTM, who had a diversified career before he assumed his position in the Tech Transfer Office of Florida State University.

Federal laboratories also have Technology Transfer Offices, and similar skills are required. Someone with very little experience can start out as a licensing associate and move up to senior licensing associate and then as a manager of a group and finally as the director of the office or beyond that as an Assistant or Associate Vice President for the Office of Technology Transfer. Typically people on these career paths would have technical degrees, even a Ph.D., and some business experience.

Opportunities Involving Patents

These include patent agents, US Patent and Trademark Office (USPTO) examiners, and patent attorneys. They all use their technical background but with additional knowledge about patent prosecution (44). The difference between a patent agent and a patent attorney is that, while they both pass the same registration exam given by the USPTO to be able to prosecute patent applications at USPTO, the patent attorney can litigate cases in court and the patent agent cannot. The patent agent is required to have an undergraduate degree in science or engineering and pass the USPTO registration examination, while the patent attorney has at least an undergraduate degree in science or engineering plus a law degree and must pass the USPTO registration exam. Patent examiners are hired by the USPTO to review patent applications and take actions on claims in patent applications. Oftentimes patent agents work in patent law firms, with a specific role to draft patent applications and prosecute applications. Some may launch their own patent prosecution firms.

In the chapter by Chao (11), he shows how IBM was able to leverage their IP portfolio to create innovation and new businesses. He also provides several examples of success stories using that strategy.

Opportunities Involving Venture Capital

Some scientists who have a strong business experience and have also been exposed to business investment can pursue careers in venture capital firms.

They are typically involved in business evaluations from both the technical and the business perspectives. They make recommendations to management on investments and depending upon the experience level, they may also be involved in business negotiations. They may move up to become partners in the firm. For those interested in such a career path, they should network and seek an internship or residency in such firms to enter the job market.

Professional Organizations

Another rewarding career path is to work for a professional organization such as the American Chemical Society (ACS), the American Association for the Advancement of Science (AAAS), and the American Institute of Chemical Engineers (AIChE). Such careers involve work with members of the organization or beyond on behalf of the professional organization. Rich, in his chapter, provides his personal experience and general lessons that can be applied to anyone interested in this career path (*13*).

Science Policy and Science Diplomacy

A policy role on the Hill is typically suited for those who have sufficient interest in policy making and political inclination. Such a path can be followed through an internship or fellowships; these include the ACS congressional fellowships (*46*) and the AAAS fellowship (*47*) in various federal agencies. These are all competitive; however, fellowships are the best options to enter this career path. Such policy fellowships can prepare someone for policy jobs at the federal, state or local level or government relations offices at universities and in public companies.

Science diplomacy entails the communication and scientific collaborations among different countries to address common problems. These may include international partnerships, collaborative research endeavors, or academic exchanges. A background or experience in government affairs and international activities is useful for someone wishing to pursue this career.

Funding Agencies

Federal funding agencies such NSF, NIH, NIST, and others hire civil service employees and also hire fixed-term Program Officers, typically from universities and some from industry. These assignments are on rotation for two to three years, and the incumbents then return to their parent institutions. The rotation option allows individuals to work at the funding agency but be paid by their parent employers with the parent insurance and pension plans. The funding agency through mutual agreements reimburses the parent employers. These Program Officers use their technical background to solicit funding proposals, arrange and coordinate the review of these proposals and make recommendations for funding.

They can also play an important role in setting the research funding agenda at the funding agencies.

In addition to government agencies, there are also several private funding sources. Many foundations provide funds for various purposes, including targeted research, publicity, or sheer philanthropy. Non-profit organizations sometimes provide funds to outsiders for activities that are usually related to their business interests. They may also hire science degree holders.

Science Journalism

For someone who has a good scientific background and enjoys writing or reporting, science journalism may be a good career option. The chapter by Baum (*14*), the editor-in-chief of *Chemical & Engineering News,* is highly recommended to anyone interested in this career path.

In addition to the opportunities shown above, there are many more career opportunities for people with chemistry training. In the literature, much has been written on alternative careers for chemists (e.g. (*48–52*)); the interested readers may want to look up these references.

It is clear that chemists can play an important role in a broad spectrum of roles, not only through traditional bench research. The key to career success lies in exploring and finding the best fit for an individual based on personal interests and aptitudes with a particular career path. This can occur at the beginning of one's career or just as well at a later stage in one's career.

Many chemists enjoy a very successful second career after switching from one sector to another. However, achieving career success and satisfaction requires identifying personal passions and strengths, often involving thoughtful planning and some strategic preparation, but most importantly, being open to exploring new pathways. It is hoped that the experiences and the advice given in the various chapters of this book will be of help to chemistry professionals in navigating their careers in the ever changing global chemistry enterprise.

Diversity and Inclusion in the Workplace

Diversity and inclusion is a major U.S. workplace issue today and will be even more important in view of globalization and the changing U.S. demographics expected in the next 50 years. Certainly this issue will impact jobs and career development in the future chemistry enterprise.

If one looks at the history of U.S. scientific and engineering development, it is clear that many of the significant discoveries and innovations have been contributed by immigrants from all over the world. More recently, discoveries and innovations in science and technology are often the results of multidisciplinary teams working together. Increasingly many businesses have recognized that perspectives and contributions from diverse team members can support both organizational and individual success. Thus, sensitivity to, and understanding of, those from diverse backgrounds is essential and contributes to success in today's diverse organizations and workforce.

In business and industry, many proponents of diversity have emphasized how it is beneficial to business interests (*53–56*). It helps to increase the talent pool for a company, better gauge customers' needs, enhance market understanding, and improve input and creativity of the workforce. This viewpoint is obviously shared by DuPont, as indicated in Gray's chapter (*28*), which has taken strong actions to promote diversity.

In academia, it has been argued that inclusion is necessary to safeguard diversity in American institutions of higher education for the sake of their students. Thus, it has been argued that diversity allows students to work and study with classmates from a diverse range of backgrounds and to enrich their experiences (*57, 58*). In his chapter, Herring (*23*) points out that research universities benefit directly from diversity—much like business organizations—because diversity (among their faculty and their students) enhances their reputational bottom lines. Tilstra, in her chapter (*24*), gives an actual example of diversity and inclusion at work at Rose-Hulman Institute of Technology. Hernandez and Watt (*22*) provide an example of a top-down approach for diversity and inclusion in the chemistry department at Georgia Tech. It is an impressive program and a good exemplar for others to follow.

Cadwalader and Bryant-Friedrich (*25*) point out that the awards process in science and technology is often not equitable. Available data show that women, industrial workers, and international workers are under-represented in award recognitions relative to their population. Their chapter suggests a number of ways to rectify this imbalance. In recent years, ACS has joined with AWIS (Association of Women in Science) and other STEM organizations and made significant efforts striving to improve awards equity without sacrifice in excellence.

In her article, Ondrechen (*26*) describes the work of the American Indian Science and Engineering Society (AISES), a national organization of Native American scientists and engineers and headquartered in Albuquerque, NM, with 3000 members. She also provides useful strategies for overcoming cultural barriers and for success in diversity and inclusion, including examples of successful synergy between Chemistry and Native America. Similar strategies can be used for other minorities. Indeed, mentorship/ apprenticeship, role models, and networks are useful means to encourage minorities to study science in schools and to ensure a steady pipeline of minority scientists for the future.

Cao in his chapter (*27*) highlights some diversity and inclusion issues relating to Asian Americans. Although the proportion of Asian Americans in science and engineering in the U.S. far exceeds the general population, the number of Asian Americans in decision making roles is woefully low in all work sectors. His chapter contains an analysis of this problem and suggests some possible solutions.

All the authors in Chapters 19-25 (*22–28*) share valuable perspectives and ongoing initiatives to raise awareness of diversity and inclusion issues impacting the workforce of the global chemistry enterprise. Since Diversity and Inclusion is such an important and contemporary topic, it is highly recommended that readers review those chapters to learn more and help spread awareness of the issues and suggestions. Visit the ACS website (59) to learn more about efforts at ACS to support Diversity and Inclusivity.

ACS Resources for Jobs and Career Development

There are many ways to develop the skills needed to thrive in the global chemistry enterprise. ACS offers a variety of valuable programs which are available to chemistry professionals (60).

Perhaps the most far-reaching single offering, the ACS Career Navigator, was launched in 2014 to assist ACS members in traversing an increasingly competitive employment landscape. This exciting new tool speaks to the concept of a career as a dynamic entity that requires continual evaluation and maintenance. Indeed, to achieve personal career goals in a rapidly changing global workforce, many individuals may want to develop new skills, branch into areas of emerging science, strengthen "soft" skills to help them become a leader in their organization, analyze market trends, or set off on an altogether new career course. Employers at the same time are seeking employees that are adaptable, able, and motivated to overcome work challenges and help their enterprise thrive. In developing the ACS Career Navigator, both of these needs were taken into consideration to ensure that chemistry professionals can achieve and exceed their individual career objectives, while helping their organizations to remain competitive and innovative.

Figure 1. Four segments of the ACS Career Navigator program. (Courtesy of the American Chemical Society.)

Interested readers are invited to check out the ACS Career Navigator website (61). As illustrated in Figure 1, it contains four distinct offerings segments that provide:

- Career services
- Leadership development training
- In-person and online professional education offerings
- Market intelligence resources

It may be useful to examine the four product quadrants further in detail. The following unique programs, products, and services all reside under the ACS Career Navigator umbrella.

Career Services

Career Pathway Workshop Series

Comprised of six separate four-hour sessions, the Career Pathway Workshop series (62) helps individuals to assess their talents, values, and interests to determine which of the career options for chemical scientists and engineers in higher education, industry, government, and entrepreneurship would be best suited to them. Workshops are primarily offered at ACS National and Regional Meetings as well as some ACS on Campus events and include:

- Finding Your Path - An overview of the major career pathways available and why each one may or may not be the right choice.
- Acing the Interview - This workshop addresses the fundamentals of successful interviewing, explaining the key differences in interviewing for different types of hiring organizations to ensure job candidates can walk into any interview with confidence..
- Working in Higher Education, Working in Industry, Working in Government, and Working for Yourself - These four sector-specific workshops provide details about each career path, opportunities present within each, and how to explore jobs in each.

Occasionally, additional workshop offerings may include:

- Electronic Tools to Enhance Your Job Hunt - Learn how to take advantage of the latest internet and information technology tools to advance your career and find a great job.
- Foreign-National Scientists: Obtaining a Job in the U.S. - Discover what needs to be done to secure a job in the United States as a non-U.S. citizen.

Career Consulting Services & Fairs

ACS also offers its members the opportunity at an ACS national meeting to attend a career fair (57) where dozens of companies are screening for hundreds of available positions. These events allow ACS members to network and explore opportunities with leading global employers of chemistry talent at a single concentrated location.

For those who cannot attend in person, a virtual career fair (63) replicates the experience allowing both domestic and international employers to participate in the International Employment Initiative (IEI) initiated last year by President Wu (4). Further career services offered by ACS under the Career Navigator include an online jobs club which meets weekly to exchange job hunting tips, network, and give support.

ACS members may also receive guidance or support from an ACS Career Consultant (*64*) on résumé reviews, interview tips, career counseling, advice on transitioning jobs, salary negotiation, and job search strategies. Career Consultants are available for individualized phone or email consultations throughout the year. Mock interview and résumé review services occur biannually at each ACS national meeting and some regional ACS meetings as well.

Leadership Development

The ACS Leadership Development System® (*65*) provides an opportunity to learn essential skills to strengthen a professional's competitive edge in today's global economy (Figure 2).

AREAS OF CORE LEADERSHIP COMPETENCY

Personal Capability	Interpersonal Skills	Focusing On Results	Setting A Clear Direction
The ACS And You	Engaging Colleagues In Dialogue	Becoming An Effective Contributor	Matching Interests With Goals
Managing Projects Effectively	Running Productive Meetings	Engaging And Motivating Volunteers	Understanding Members' Interest
Fostering Innovation	Coaching And Feedback	Collaborating Across Boundaries	Leading Change
Leading Without Authority	Developing Communication Strategies	Succession Planning	Strategic Planning
Extraordinary Leader — 8 Hour Course			

← Character Competency →

▇▇▇ On-line Courses   ☐ 4-Hour Facilitated Courses

Figure 2. Courses available in the ACS Leadership Development System. (Courtesy of the American Chemical Society.)

The suite of courses available provides an opportunity for both face-to-face as well as online learnings around the four main competencies of: personal capabilities, interpersonal skills, focusing on results, and setting a clear direction. These characteristics have been found to differentiate outstanding leaders in corporate and other organizational environments. They are offered to ACS members and leaders yearly at ACS national meetings, ACS Leadership Institute, and specially organized local and regional events.

Professional Education

ACS Professional Education is the recognized source for vital training needed to keep skills at their peak. Created in 1965 by the ACS Department of Education Activities, offered courses are designed to meet the education needs of scientists and engineers from varying backgrounds, nationalities, employment situations, and experience levels.

ACS Professional Education (66) offers a wealth of professional development courses delivered through multimodal platforms: from on-site learning, to online delivery, to the newest online learning environment, Sci-Mind™. The courses have the essential content and methodology to ensure learners and their employers meet their goals around both technical and non-technical essential skills. Courses are taught by experienced instructors who are leaders in their respective fields, cover a wide array of chemistry disciplines including most major subdisciplines of chemistry and emerging frontiers, and serve scientists from over 65 countries.

Approximately 50 different topics are offered each year at the ACS national meetings, through public courses arranged throughout the United States, customized courses delivered to corporate partners, and online and on-demand offerings delivered virtually regardless of learner geography.

Market Intelligence

ACS conducts annual and 5-year surveys to learn about salary, demographic, and general employment trends. Reports based on these surveys include: salary data and employment information for chemists and chemical engineers in the United States, overall employment trends in the chemistry enterprise, and employment guidance in areas of emerging technologies and employment opportunities. ACS members receive unrestricted access to the survey results and data (67). The salary calculator, currently in its 12th edition, contains salary data by career choice, degree level, geography, and specialty among other options. An ACS member can enter a real or hypothetical case with specific information on degree level, specialization, and geographical region and receive probable salaries based on the information in the database. This valuable information has benefitted many ACS members in their job searches and salary negotiations after obtaining a job offer.

A Wealth of Other Resources

Additional useful resources that ACS offers which reside outside the ACS Career Navigator include:

- ACS local section meetings provide a great opportunity to network, and develop relationships which increase one's visibility and help one to discover hidden job opportunities.
- ACS Webinars (www.acswebinars.org) on a variety of topics expand one's knowledge about career skills and other valuable information.
- Many universities host ACS two-day workshops on Preparing for Life After Graduate School (PfLAGS). These workshops examine career options for Ph.D. chemists, critical non-technical skills, employment

opportunities, and preparation for academic positions, while providing individualized assistance at putting a strong career-search plan in place.
- Another useful program is ACS on Campus (http://acsoncampus.acs.org/), which brings experts to universities around the world. Some components of these day-long events include networking science cafes, basics in scholarly publishing, navigating science technical writing, grant proposal writing, and entrepreneurship.
- There are opportunities to succeed through a willingness to work internationally. The ACS International Center™ (www.acs.org/ic) is the premier clearinghouse of information on educational and career exchange programs worldwide, regardless of the country. There are many educational and exchange programs that support the development of competency in working with other cultures.
- Small companies and startups are a source of many opportunities and, according to the latest trends, offer the most new jobs. ACS provides support to startup companies founded by ACS members through the ACS Entrepreneurial Resources Center (www.acs.org/ei). Resources such as connections, funding, managerial and business development advice, and professional services are provided to the most promising new firms. Other resources for entrepreneurs include programming at National and Regional ACS Meetings, and a series of ACS webinars (available for ACS members through ACS Presentations on Demand).

In addition, the ACS has 32 Technical Divisions that provide programming for ACS national meetings. Each technical division is field specific and provides members a forum for networking and collaboration within their field (*68*). Moreover, some divisions keep track of new job openings and even permit their members to post their CV's on their websites. It is highly advisable for a chemistry professional to join one or more ACS divisions in the fields where he or she is professionally active.

- For people interested in business, other sources of information at ACS include Division of Small Chemical Businesses (SChB), Division of Industrial and Engineering Chemistry (I&EC), Division of Business Development and Management (BMGT), and Division of Professional Relations (PROF). SChB may be particularly relevant to people interested in starting a new business.
- For people interested in a career in law and/or patents, a resource for more information is the ACS Division of Chemistry and the Law (CHAL). It holds receptions at national ACS meetings which provide excellent networking opportunities to meet and talk with chemists who have worked with IP or become patent attorneys.

Beyond ACS, there are career resources available from universities, employers, and online. There are innovative academic programs, new approaches to employee professional development, community networks, and massive online open courses (MOOCs) on relevant topics.

Conclusions

On balance, there are more opportunities than challenges in the global chemistry enterprise of the future. Thus, the latest data show U.S. chemistry unemployment heading downward, along with a decrease in part-time vs. full-time work. Moreover, chemistry continues to have a vital role in addressing the world's challenges. It is important for chemistry professionals to remain aware of the global trends. They need to be flexible, keep on gaining new knowledge, and stay active professionally. Chemistry professionals and students also need to be acquainted with the full range of career options open to them – chemistry jobs, chemistry related jobs, science related jobs, science inspired jobs, and non-traditional career paths. For the right individual and circumstances, international work experiences can be enriching and valuable.

The ACS has many resources that can help its members. Many of these resources have been summarized in this article. Hopefully this information will be useful to students and working professionals alike as they chart their career paths in the ever-changing global chemistry enterprise.

Acknowledgments

The authors would like to thank many individuals for their contributions to this article. Special mention should be made of Steven Meyers, who provided information on ACS resources on jobs and career development, and Elizabeth McGaha and Gareth Edwards for supplying the ACS survey data. Steve Meyers also critiqued the entire article, and helpful comments were received from Chanel Fortier and Alvin Bopp. Thanks are due to Max Saffell, David Harwell, and Joy Titus-Young for support. Thanks are also due to the authors of all the chapters in this book for their insights and perspectives.

Mention of trade names or commercial products in this publication is solely for the purpose of providing specific information and does not imply recommendation or endorsement by the U.S. Department of Agriculture. USDA is an equal opportunity provider and employer.

References

1. Global Chemicals Outlook, United Nations Environment Programme, 2012. http://www.unep.org/pdf/GCO_Synthesis%20Report_CBDTIE_UNEP_September5_2012.pdf.
2. ACS Business Department. Lackluster year for chemical output. *Chem. Eng. News* **2013**, *91* (26), 41.
3. Guide to the Business of Chemistry – 2013. American Chemistry Council. www.americanchemistry.com/business-of-chemistry-summary.
4. Partners for Progress and Prosperity in the Global Chemistry Enterprise, *Vision 2025: How to Succeed in the Global Chemistry Enterprise*. Cheng, H. N.; Shah, S.; Wu, M. L. Eds.; ACS Symposium Series 1157; American Chemical Society, Washington, DC, 2014.

5. Lechleiter, J. C. Why Chemistry Still Matters. *Careers, Entrepreneurship, and Diversity: Challenges and Opportunities in the Global Chemistry Enterprise*; ACS Symposium Series; American Chemical Society: Washington, DC, 2014; Vol. 1169; Chapter 2.
6. Laurence, J. S. Academic Opportunities and Considerations for a Successful Career in an Evolving Institution. *Careers, Entrepreneurship, and Diversity: Challenges and Opportunities*; ACS Symposium Series; American Chemical Society: Washington, DC, 2014; Vol. 1169; Chapter 3.
7. Shulman, J. I. Industrial Careers: Obtaining a Job and Succeeding in Industry. *Careers, Entrepreneurship, and Diversity: Challenges and Opportunities*; ACS Symposium Series; American Chemical Society: Washington, DC, 2014; Vol. 1169; Chapter 4.
8. Omberg, K. M. Career Opportunities in the Government Sector. *Careers, Entrepreneurship, and Diversity: Challenges and Opportunities*; ACS Symposium Series; American Chemical Society: Washington, DC, 2014; Vol. 1169; Chapter 5.
9. Frishberg, M. D. From the Lab to a Career in Business Development: An Unexpected Career Transition. *Careers, Entrepreneurship, and Diversity: Challenges and Opportunities*; ACS Symposium Series; American Chemical Society: Washington, DC, 2014; Vol. 1169; Chapter 6.
10. Rodriguez, G. Connecting with a Most Powerful Ally of Science. *Careers, Entrepreneurship, and Diversity: Challenges and Opportunities*; ACS Symposium Series; American Chemical Society: Washington, DC, 2014; Vol. 1169; Chapter 7.
11. Chao, J. L. Alternative Chemistry Careers: Using Patent Assets for High Technology Commercial Innovation. *Careers, Entrepreneurship, and Diversity: Challenges and Opportunities*; ACS Symposium Series; American Chemical Society: Washington, DC, 2014; Vol. 1169; Chapter 8.
12. Fraser, J. A. And Then the Wind Changed. *Careers, Entrepreneurship, and Diversity: Challenges and Opportunities*; ACS Symposium Series; American Chemical Society: Washington, DC, 2014; Vol. 1169; Chapter 9.
13. Rich, R. H. A Chemist Strategizing for Chemists: Careers in Societies and Associations. *Careers, Entrepreneurship, and Diversity: Challenges and Opportunities*; ACS Symposium Series; American Chemical Society: Washington, DC, 2014; Vol. 1169; Chapter 10.
14. Baum, R. M. From Med-School Drop-Out to Editor-in-Chief of C&EN. *Careers, Entrepreneurship, and Diversity: Challenges and Opportunities*; ACS Symposium Series; American Chemical Society: Washington, DC, 2014; Vol. 1169; Chapter 11.
15. Beattie, T. R. Consulting for Seniors: Having It Your Way. *Careers, Entrepreneurship, and Diversity: Challenges and Opportunities*; ACS Symposium Series; American Chemical Society: Washington, DC, 2014; Vol. 1169; Chapter 12.
16. Brown, R. E.; Fraser, E. I.; Polk, K. J.; Harwell, D. E. The American Chemical Society (ACS) Entrepreneurial Initiative. *Careers, Entrepreneurship, and Diversity: Challenges and Opportunities*; ACS

Symposium Series; American Chemical Society: Washington, DC, 2014; Vol. 1169; Chapter 13.
17. Shah, S. Creativity, Innovation and the Entrepreneurship Nexus. *Careers, Entrepreneurship, and Diversity: Challenges and Opportunities*; ACS Symposium Series; American Chemical Society: Washington, DC, 2014; Vol. 1169; Chapter 14.
18. Confalone, P. N. Innovation and Entrepreneurship in the Chemical Enterprise. *Careers, Entrepreneurship, and Diversity: Challenges and Opportunities*; ACS Symposium Series; American Chemical Society: Washington, DC, 2014; Vol. 1169; Chapter 15.
19. Ornstein, J. M.; Zhou, H. Venture Capitalist Planning Is Irrelevant to Successful Entrepreneurship in Chemicals, Materials, and Cleantech. *Careers, Entrepreneurship, and Diversity: Challenges and Opportunities*; ACS Symposium Series; American Chemical Society: Washington, DC, 2014; Vol. 1169; Chapter 16.
20. Lawrence, N. J.; Brewer, J.; Kruse, A. A Tale of Academic Innovation to Job Creation in Rare Earth Extraction and Separation. *Careers, Entrepreneurship, and Diversity: Challenges and Opportunities*; ACS Symposium Series; American Chemical Society: Washington, DC, 2014; Vol. 1169; Chapter 17.
21. Vercellotti, S. V.; Vercellotti, J. R. Technical Entrepreneurship Serving Industry: A Personal Story. *Careers, Entrepreneurship, and Diversity: Challenges and Opportunities*; ACS Symposium Series; American Chemical Society: Washington, DC, 2014; Vol. 1169; Chapter 18.
22. Hernandez, R.; Watt, S. A Top-Down Approach for Diversity and Inclusion in Chemistry Departments. *Careers, Entrepreneurship, and Diversity: Challenges and Opportunities*; ACS Symposium Series; American Chemical Society: Washington, DC, 2014; Vol. 1169; Chapter 19.
23. Herring, C. Diversity and Departmental Rankings in Chemistry. *Careers, Entrepreneurship, and Diversity: Challenges and Opportunities*; ACS Symposium Series; American Chemical Society: Washington, DC, 2014; Vol. 1169; Chapter 20.
24. Tilstra, L. Promoting Diversity and Inclusivity at Rose-Hulman Institute of Technology. *Careers, Entrepreneurship, and Diversity: Challenges and Opportunities*; ACS Symposium Series; American Chemical Society: Washington, DC, 2014; Vol. 1169; Chapter 21.
25. Cadwalader, E. L.; Bryant-Friedrich, A. C. Improving Transparency and Equity in Scholarly Recognition by Scientific Societies. *Careers, Entrepreneurship, and Diversity: Challenges and Opportunities*; ACS Symposium Series; American Chemical Society: Washington, DC, 2014; Vol. 1169; Chapter 22.
26. Ondrechen, M. J. American Indian Science and Engineering Society: Building a Successful Model for Diversity and Inclusion. *Careers, Entrepreneurship, and Diversity: Challenges and Opportunities*; ACS Symposium Series; American Chemical Society: Washington, DC, 2014; Vol. 1169; Chapter 23.

27. Cao, G. Career Opportunity Challenges for Asian American Chemistry Professionals and Scientists. *Careers, Entrepreneurship, and Diversity: Challenges and Opportunities*; ACS Symposium Series; American Chemical Society: Washington, DC, 2014; Vol. 1169; Chapter 24.
28. Gray, T. Q. Diversity and inclusion from a global perspective. *Careers, Entrepreneurship, and Diversity: Challenges and Opportunities*; ACS Symposium Series; American Chemical Society: Washington, DC, 2014; Vol. 1169; Chapter 25.
29. Rovner, S. L. 2013 Salaries & Employment. *Chem. Eng. News* **2013**, *91* (38), 9–13.
30. ACS Comprehensive Salary and Employment Survey, 2013.
31. http://www.aacu.org/leap/documents/2013_EmployerSurvey.pdf.
32. http://www.iie.org/~/media/Files/Services/ProjectAtlas/Project-Atlas-Trends-and-Global-Data-2013.ashx.
33. http://teach.com/why-teach/the-demand-for-great-teachers.
34. Community College funding priorities: http://www.aacc.nche.edu/Advocacy/Pages/federal_priorities.aspx.
35. Fain, P. *Federal Spending that Works, Inside Higher Education*, May 14, 2013; http://www.insidehighered.com/news/2013/05/14/labor-department-grants-may-be-paying-community-colleges-and-students.
36. The STEM Workforce Challenge: the Role of the Public Workforce System in a National Solution for a Competitive Science, Technology, Engineering, and Mathematics (STEM) Workforce (prepared for the U.S. Department of Labor: *Employment and Training Administration by Jobs for the Future*), 2007.
37. Nash, J. A.; Wright, D. A. Profile of the Chief Research Officer at Major Research Universities in the United States and Examination of the Current Pathways to the Position. *J. Research Administration* **2013**, *2*, 74–93.
38. Southwest Research Institute is an example of a free standing research institute: http://www.swri.org/
39. Helfand, J.Sadeghi, A.Talan, D. Employment dynamics: small and large firms over the business cycle. *Monthly Labor Reviews*; March 2007; pp 40–50.
40. http://www.strem.com/about/.
41. http://www.gore.com/en_xx/aboutus/timeline/index.html.
42. http://www.acs.org/content/acs/en/careers/profdev/resourcecenter.html.
43. http://en.wikipedia.org/wiki/List_of_federally_funded_research_and_development_centers.
44. To view additional information about patent examiners: http://careers.uspto.gov/Pages/PEPositions/.
45. Association of the University Technology Transfer Managers (AUTM): http://www.autm.net/Home.htm.
46. ACS Congressional Fellowship Program: http://www.acs.org/content/acs/en/policy/policyfellowships.html.
47. The American Association for the Advancement of Science Fellowship Program: http://www.aaas.org/page/fellowships.

48. Balbes, L. M. *Nontraditional Careers for Chemists: New Formulas in Chemistry*; Oxford University Press: New York, 2007.
49. Owens, F.; Uhler, R.; Marasco, C. A. *Careers for Chemists: A World outside the Lab*; American Chemical Society: Washington, DC, 1997.
50. http://cenblog.org/just-another-electron-pusher/2012/10/so-many-nontraditional-chemistry-careers/.
51. http://www.acs.org/content/acs/en/careers/college-to-career/chemistry-careers.html.
52. http://www.ncl.ac.uk/careers/explore/occupations/Scienceoutsidethelab.php.
53. Hewlett, S. A.; Marshall, M.; Sherbin, L. How Diversity Can Drive Innovation. *Harvard Business Rev.*, December 2013. http://hbr.org/2013/12/how-diversity-can-drive-innovation/ar/1.
54. Anon. Global Diversity and Inclusion: Fostering Innovation Through a Diverse Workforce. *Forbes Insight*, 2011. http://images.forbes.com/forbesinsights/StudyPDFs/Innovation_Through_Diversity.pdf.
55. Robinson, M.; Pfeffer, C.; Buccigrossi, J. *Business Case for Inclusion and Engagement*. wetWare, Inc.: Rochester, NY, 2003. http://workforcediversitynetwork.com/docs/business_case_3.pdf.
56. Anon. *What is the business case for diversity?* http://www.workforce.com/articles/20086-whats-the-business-case-for-diversity.
57. Gurin, P.; Nagda, B. A.; Lopez, G. E. The Benefits of Diversity in Education for Democratic Citizenship. *J. Social Issues* **2004**, *60* (1), 17–34.
58. *Does Diversity Make a Difference? Three Research Studies on Diversity in College Classrooms*. American Council on Education and American Association of University Professors: Washington, DC, 2000.
59. http://www.acs.org/content/acs/en/membership-and-networks/acs/welcoming/diversity.html.
60. ACS Career Navigator/Professional Resources. http://www.acs.org/content/acs/en/membership-and-networks/member-handbook/continuing-education-career-resources.html.
61. http://www.acs.org/content/acs/en/careers.html.
62. http://www.careerfair.acs.org/agenda/onsite-workshops.
63. http://www.careerfair.acs.org/.
64. http://www.acs.org/content/acs/en/careers/ccp.html.
65. http://www.acs.org/content/acs/en/careers/profdev/leadership.html.
66. http://www.acs.org/content/acs/en/careers/profdev/continuing.html.
67. http://www.acs.org/content/acs/en/careers/salaries.html.
68. http://www.acs.org/content/acs/en/membership-and-networks/td.html.

Career Advancement Opportunities

Chapter 2

Why Chemistry Still Matters

John C. Lechleiter*

Chairman, President, and Chief Executive Officer, Eli Lilly and Company, 893 S. Delaware Street, Indianapolis, Indiana 46285
*E-mail: jcl@lilly.com

Chemistry is a solid foundation for building a career, and chemists can prepare for career paths that lead to multiple roles inside or outside the lab. Chemistry continues to play a central role in maintaining American competitiveness and solving today's global challenges, including efforts to discover and develop innovative medicines that address unmet patient needs. Chemists can participate in an exciting period for drug discovery, with extraordinary opportunities for collaboration among industry, academia, and government labs. In addition, the knowledge and skills developed in the study of chemistry are essential to many other areas across the public and private sectors. To realize the full potential of their training, chemists need to develop attributes that are essential to any successful career, such as the ability to express oneself clearly and awareness and appreciation of the external environment.

Since joining Lilly nearly 34 years ago as an organic chemist in the process research and development division, I have taken on a range of responsibilities within the company, and for the past five years have been privileged to serve as CEO.

This article will discuss, from the perspective of my current role, how chemistry continues to be at the core of our efforts to discover and develop innovative pharmaceuticals that address unmet patient needs.

It will also discuss how the knowledge and skills that all of us have developed in the study of *chemistry* are essential to many *other* areas across the public and

© 2014 American Chemical Society

private sectors – and how chemists can prepare for career paths that can lead to multiple roles inside or outside the lab.

The Central Role of Chemistry in Meeting Global Challenges

The first reason why chemistry still matters, and why chemistry is a solid foundation for building a career, is that chemistry and the physical and biological sciences will continue to play a central role in maintaining American competitiveness and solving the global challenges we face today – challenges that include protecting the environment, developing new energy sources, feeding a hungry world, and conquering the scourge of disease in both developed and developing nations. This last challenge will be the focus here.

The need for new medicines cannot be overstated. Around the world, we face disturbing trends in the rise of diabetes, cancer, and cardiovascular disease. Our populations are aging. And emerging global economies will create unprecedented demand for better care and access to pharmaceuticals.

Yet, it is no secret that the pharmaceutical industry is in the midst of a period of almost unprecedented change. There are many forces acting on the industry: rising costs and reduced productivity in R&D, difficulties with long development times and looming patent expirations, concerns over drug safety, and increasing pressure on pharmaceutical prices.

At the same time, we are witnessing extraordinary scientific breakthroughs that are providing invaluable insights into human biology and disease pathways, suggesting new opportunities to improve on existing therapies as well as to address needs that are as yet unmet.

Indeed, I have argued that we are at the beginning of what will come to be known as the Biomedical Century, an era when scientific advances will drive gains in human health and longevity surpassing even those of the past 100 years.

So how do we seize the opportunity before us?

The answer, I believe, is embedded in the words of Eli Lilly, the grandson of our founder and one of our company's greatest leaders, who stated in 1946: "Research is the heart of the business, the soul of the enterprise." These words transcend time and are a reminder to our scientists of the pivotal role they play in bringing health and hope to patients.

And while a number of our peer companies have made significant cuts to their R&D budgets and staffing, Lilly remains firmly committed to innovation. Despite significant financial pressures during the current period when we face patent expirations for several of our largest products, we have sustained investment in R&D to take advantage of the opportunities before us and generate sustained growth in the long term.

More than 7,600 scientists, physicians, engineers, statisticians, and other talented individuals are working at 11 Lilly research and development sites around the world on new disease targets in neuroscience, cancer, diabetes, autoimmunity, cardiovascular disease, and other areas.

Let me emphasize that chemistry has been, and will remain, central to creating new medicines and delivering better treatment options to patients. I always hearken back to this simple truth: Chemistry enables the creation and characterization of truly new matter; that is, heretofore unknown materials with new and unique chemical, physical, and biological properties. It is a powerfully enabling discipline that has already transformed human life in many ways and produces still the currency of our innovation enterprise.

The discipline of synthetic chemistry, in particular, supports many activities along the continuum of drug discovery. At its core, our ability to understand and master molecular structure, function, and properties through synthesis is a key source of innovation.

- In the early stages of drug discovery, we focus on enabling chemistry that allows us to test diverse molecular hypotheses as part of iterative structure-activity studies to create high-quality clinical candidates.
- In the later phases of discovery, synthetic chemistry focuses on the demands of strategic material supply and the scale-up work necessary to support advanced pre-clinical and eventually clinical studies and large-scale manufacture.
- More recently, our chemists have been collaborating in new ways with our protein engineers to attach small molecules to antibodies and to alter and improve the utility of other large molecules through chemical modification.

All along this path we engage companion disciplines such as the analytical sciences to purify and characterize new molecules.

While this general framework has remained fairly constant over the past several decades, one thing has not: The structural complexity of our molecules has increased significantly with the growing sophistication of our work and the diversity of therapeutic modalities that we can now design through the use of computational and biophysical technologies.

We are exploring concepts in our laboratories today that 10 years ago were unimaginable. Synthetic chemistry plays an essential role in transforming these concepts into chemical matter and ultimately bringing the highest-quality solution to the patient.

Broad Opportunities for Chemists To Participate in Biomedical Research

In addition, we are working in new ways with chemists and other scientists outside our organization, in academia, government labs, and small biotech companies. Lilly is involved with more than 50 public-private partnerships and consortia, engaging with other leading experts around the world. In fact, we are seeing this kind of collaboration across the biopharmaceutical industry, as companies seek access to a wider network for molecules and talent.

Let me offer a few examples from our own firm.

The first is our Open Innovation Drug Discovery initiative, launched in 2009, which allows us to collaborate with scientists in academia and small biotechs and opens the door for accessing promising molecules from around the world. Through this program, Lilly carries out screening tests – free of charge – on compounds submitted by outside researchers. In return, we retain first rights to negotiate an agreement with them. If no such agreement results, external researchers receive no-strings-attached ownership of the data report from Lilly to use as they see fit in publications, grant proposals, or further research.

As of mid-2013, more than 350 universities, research institutes, and small biotechs representing 32 countries were affiliated with the program, and we have received to date some 185,000 chemical structures uploaded into our database for evaluation. We have entered 11 collaborations, with five now completed.

Another way we are engaging with scientists outside our walls is through Chorus, a small, cross-disciplinary group of Lilly scientists established in 2002 that designs and oversees early-stage development work through a network of organizations outside Lilly. More recently, we have worked with venture capitalists to create virtual project-focused companies that are taking molecules from pre-clinical development to clinical proof of concept. If proof-of-concept studies are positive, the molecules are made available for licensing to biopharmaceutical companies, including Lilly.

We are also involved in a range of activities with the National Institutes of Health. A unique collaboration with the National Center for Advancing Translational Science, for example, allows NIH scientists to perform chemical reactions in Lilly's one-of-a-kind Automated Synthesis Lab in Indianapolis from their computers on the NIH campus.

Lilly's automated laboratory functions as a remotely guided pair of hands for the NIH scientists, providing them with real-time visual and analytical feedback on their reaction, just as if it were being performed in their own fume hood. In this way, we share the molecules that are synthesized for our independent use and at the same time expand the diversity of synthetic chemistries that can be performed in an automated fashion. This is certainly taking public-private partnerships in a new direction!

Lilly is also part of a major new initiative here in Indiana, focused on collaboration between Indiana's life sciences industry and its major research universities. In May, leaders of the state's life sciences community joined with Governor Pence to unveil the Indiana Biosciences Research Institute – the first industry-led biosciences research institute in the nation. The state legislature has approved $25 million in start-up funding, and we are working to raise additional capital, to identify and recruit a CEO, and to develop the rules of engagement for intellectual property and commercialization made by all partners and the institute.

This is, without a doubt, one of the most exciting periods for chemists to be involved in drug discovery, with a deep pool of new scientific knowledge at our disposal, new tools that we can apply to our efforts, and extraordinary opportunities for collaboration among industry, academia, and government labs.

Building on Skills Developed in the Study of Chemistry

Another reason why chemistry still matters is that the discipline of chemistry develops knowledge and skills that are critically important, not only to pharmaceutical R&D, but also to a whole range of roles in our complex and rapidly changing world.

I am representative of many thousands of people who have advanced degrees in chemistry and yet today are working outside the lab. I am sure we can all point to things we learned in our academic and professional training as chemists that have served us well in our current roles. Let me suggest just a few examples:

As scientists by training, we carry with us the ability to think critically. We have had to develop the skills necessary to identify, analyze, and solve problems. This kind of problem-solving expertise, along with the initiative and persistence that entails, will always be in hot demand in any enterprise.

Chemists also understand the importance of data in guiding decision making. In particular, chemists are particularly adept at transforming data into knowledge and using data to generate new hypotheses. This keeps the gears of innovation turning. At a time when "Big Data" is the latest buzzword, our training has given us a leg up in dealing with the ever-growing mass of information that confronts modern organizations.

In addition, the pursuit of chemistry teaches us to appreciate and master complexity, a quality that helps us avoid oversimplifying complex matters, on the one hand, or being overwhelmed by them, on the other. Certainly, my study of chemistry provides me the understanding of science that greatly assists me in leading a science-based company. But it also provides a way of thinking that helps make sense of many of the other complex aspects of a modern corporation.

In sum, our training as chemists helps prepare us for a wide range of careers outside the lab.

At the same time, however, it is important for us to be aware of where we might need to stretch beyond the dispositions that brought us to scientific work, and the tendencies reinforced by the scientific training we have received, to be most effective in our organizations and in our careers.

Again, let me explore this issue from my perspective in the biopharmaceutical industry.

Developing medicines is a long, complex process – a truly massive effort requiring an investment of hundreds of millions of dollars and lasting a decade or more. It is not the undertaking of any one person, and it demands a wide range of expertise as a potential medicine moves through the development process.

The path of advancement in academia, along with the organization of academic departments and of groups within departments, can encourage a deep but often narrow focus, and an emphasis on individual achievement over collaboration. This sort of approach can be valuable, certainly. It is at the heart of the academic model and has led to the preeminence of American research universities on the world scene.

In our industry, we definitely need scientists who are deeply knowledgeable in their own disciplines. But they must also have sufficient knowledge of other disciplines to collaborate effectively with their colleagues. Indeed, with the explosion of knowledge and communication technologies, effective research scientists must be able to seek and connect knowledge across a broad network both inside and outside the company.

They need to be able to work with other people from a wide range of disciplines and backgrounds, and also curious enough to want to know what is going on in the lab across the hall. With this spirit of openness, they can begin to master the paradox of being both a strong team player and individual contributor, to achieve outcomes greater than any single individual can accomplish.

These qualities – depth of knowledge, breadth of knowledge, and an ability to communicate and to work with others in increasingly diverse settings – are essential not just for scientists, but for the whole gamut of responsibilities throughout our organization, particularly for those in leadership roles.

At Lilly, when we recruit and assess talent across our enterprise, we look for people who are – to use the terminology developed by Michael Lombardo and Robert Eichenger – "learning agile (*1*)." These are people who learn readily from their experience and can apply that knowledge in new and difficult situations to perform at a high level. Eichenger and Lombardo found that learning agility is also highly correlated with leadership potential.

We have distilled 12 factors we think are critical to learning agility. For example, we seek people who are insatiably curious, people with foresight, who can identify root causes and underlying patterns and trends, who draw upon broad sources of knowledge, who are open to others and their ideas, who can take the heat and lead change.

And let me add two qualities that are absolutely essential to any successful career today.

The first is the ability to express yourself clearly, to be able to communicate complex and sometimes contentious issues in ways that make sense not only to you, but to the people you are attempting to reach. And while Twitter has its place (in fact, Lilly has an established presence in social media), we all do need to be able to express ideas that require more than 140 characters!

The second essential quality is an external focus, an awareness and appreciation of the wider environment in which your organization operates. The world is the opposite of a controlled experiment, and success requires the ability to identify and adjust to rapidly changing variables. As in the work of drug discovery and development, success seldom comes from the perfect plan executed perfectly. It is more often the result of dealing effectively with anticipated and unanticipated challenges, drawing on knowledge and resources from the world over.

Conclusion: Why Chemistry Is Still a Solid Choice

In conclusion, pursuing education and professional development in chemistry will continue to be a solid foundation for career advancement in the Biomedical Century ahead, both *in* and *out* of the lab.

And yet, I know that many of us chose to pursue degrees in chemistry for another reason altogether: It was our passion! It was fun! It was rewarding!

Like many others, I developed a passion for chemistry at an early age. Through science fairs, undergraduate research, programs, and a series of excellent teachers – themselves passionate about chemistry – I grew determined to pursue chemistry as a career.

Yet, if all I looked at was the dismal job market for chemists in 1975, the year I entered graduate school in chemistry, I might have changed course. Fortunately, I did not, and by the time I joined Lilly in 1979, the picture was completely different.

In the three decades since, we have seen more ups and downs in the field, but what has *not* changed is the fundamental importance of chemistry, and the critical role that individuals trained in this discipline can play in meeting the challenges of our world. Yes, *chemistry still matters* – to my life and my job, to the work of our company, and to the future of our country and our world.

We will continue to need people who not only have an understanding of science, but also possess the skills and attributes necessary to apply that understanding in a wide range of organizations -- people who are not only experts in chemistry, but excited about it, people who know that chemistry still matters.

I hope that these reflections provide some affirmation for experienced professionals, encouragement to those who are still finding their way, and, for those seeking careers in chemistry, a glimpse of the exciting possibilities that await you.

Acknowledgments

This article is based on an address to the Presidential Symposium on Career Advancement, American Chemical Society National Meeting, September 9, 2013, Indianapolis, IN.

References

1. Lombardo, M. M.; Eichinger, R. W. *Human Resource Management* **2000**, *39* (4), 321–29.

Chapter 3

Academic Opportunities and Considerations for a Successful Career in an Evolving Institution

Jennifer S. Laurence[*]

Department of Pharmaceutical Chemistry, University of Kansas,
2030 Becker Drive, Lawrence, Kansas 66047-1620
*E-mail: laurencj@ku.edu

The goal of this chapter is to provide a framework for understanding what it takes to secure a faculty position, successfully advance in this career path and plan for a sustainable career. Academic scientists advance and disseminate knowledge through their research, teaching and service activities. These pursuits build on the scientific foundation provided during pre- and post-doctoral training but also encompass an increasingly diverse set of roles, opportunities and responsibilities. This chapter provides an introduction to various aspects of the professorial career path. In addition, a case will be made for the importance of individual and institutional interconnection with partners outside the university, as these relationships are critical to effective translation of and, increasingly, funding for research. Expectations for both traditional and emerging ideas about faculty roles and responsibilities will be discussed.

Introduction

The path of an academic traditionally has been a highly individual undertaking, so much so that colorful sayings have emerged to describe how challenging the job of being a leader within an academic structure may be. Getting a group of academics to achieve a common goal has been described as herding cats or nailing jello to a tree, which certainly invokes a vivid image, but also conveys a valid point. With this in mind, it should be no surprise that a career in academe best suits those who are strongly self-directed, highly self-motivated

© 2014 American Chemical Society

and thrive in a constantly evolving environment. Academic scientist also must have exceptional focus and ambition directed toward accomplishing a unique goal within their field of study. In keeping with this mode, if one asks a group of faculty to state the mission of a university a diverse array of reponses are likely to be provided. Most, however, would not disagree that the academic mission is "to create, advance, and disseminate knowledge and inspire excellence", a set of ideals often reflected in formal university mission statements. The associated activities through which these responsibilities are executed usually include independent research; teaching at the baccalaureate and post-graduate levels; service to the department, to the university, and to the scientific profession at large and, in some cases, to the broader community.

In fundamental concept the academic mission has been steadfast for many centuries and remains so still. Nonetheless, the world in which we live continues to undergo significant transformations that impact the way the academic mission is carried out. Some rather profound changes have occurred in the past several years that affect the academic landscape, and some of the issues that warrant consideration are listed here:

1) Research funding has been scarce and in decline. It has changed forms as well and shifted toward an emphasis on translational as opposed to fundamental research. Certainly U.S. government funding has been very competitive due to federal budgetary constraints. Industrial funding, albeit less traditional for academics to pursue, has been a source of support in certain fields and has become even more difficult to obtain as the U.S. companies attempt to adapt to changing economic factors and the constant demands on their bottom line have increased in recent years.

2) Traditional academia is considered the fountainhead of fundamental research. There is, however, increasing pressure to translate fundamental research to applications and commercialize technology. The training academics receive typically prepares them well to conduct research, but it provides little experience or knowledge of the processes involved in moving research into a development phase or the commercial market. The priorities of the two differ and many important considerations for success in these arenas are unknown to the academic. Patent applications and intellectual property are important factors to consider in such situations, which imposes time and financial constraints on faculty and their institution. It also creates potential competing interests, such as the need to delay publication for the sake of a patent. The two outcomes are not weighted equally in the academic review process, making this is an important consideration during the early career period.

3) In an effort to maintain funding, some professors have placed more research emphasis on applied or product-directed areas. Some professors have started their own companies, and a few have been successful as entrepreneurs. This can be major time commitment and result in conflicts of interest. Because a professor's primary responsibility is to the university, the process and implications of the decision to participate in a business are monitored by representatives of the university.

4) It has been said that there are no small problems left to solve. The implication of this is that more sophisticated approaches are required to understand complex processes and overcome challenging problems. Because one person can have only limited experience and knowledge, many people have found collaboration to be advantageous as they attempt to accelerate their research, improve their work efficiency, and expand their R&D capabilities. Indeed, companies have realized for a long time that teamwork is essential to support an effective and efficient industrial development process.
5) Much to the dismay of many professors, jobs for chemistry graduates have been difficult to find in the past several years. As has been demonstrated in the tech sector, start up companies are an engine for job creation. The parameters of applying this to the chemical sciences obviously differ, but significant efforts are being made in a variety of forms to support scientific entrepreneurship.
6) There is increasingly a shift towards on-line education. The so-called Massive Open Online Courses (MOOC) have become popular. Many universities have embraced on-line education and provided different types of options. Specialized schools that only provide on-line education have also sprung up, with varying degrees of success.
7) The rapid electronic communication that is fast, cheap, and efficient has changed the pace of research and the manner in which it is done. Interactions between professors and students are increasingly being conducted on-line. Electronic communication also facilitates collaborations across geographical or national boundaries.

All these changes have created many challenges and opportunities for academia. A person working in or contemplating a career in academia needs to be cognizant of these changes and adapt to the situations in order to thrive and accomplish the core mission of generating and disseminating knowledge. Some instructive books have been published previously (1–3). In this chapter, I use my own experience to provide examples of how to deal with certain types of challenges and opportunities one may face in an academic career.

My Career Path

Individual career paths differ substantially. Some paths result from early knowledge of one's passion and the development of long-term plans to achieve a specific goal, while others are driven by general curiosity and find interest in a broad diversity of experiences and fields of study, which may coalesce into the decision to pursue a particular direction as result of a significant event. My path belongs to the latter category. I began my undergraduate studies interested in many subjects, ranging from international studies to medicine, and ultimately received a B.A. degree in history from Miami University in Oxford, Ohio. While most do not cross into chemistry from such distant fields for obvious reasons, the experiences and training provided in the course of these studies offered some significant advantages, including the opportunity to develop and mature my writing, communication and logic skills. I spent time in a study-abroad program in Israel, which was an incredibly rich experience. It brought the scientific process to life; as I worked with a team of experts from an array of fields, I learned about the systematic approach to field archaeology and how chemical science is applied to examine and understand artifacts. I found myself particularly drawn to the process of discovery and analytical validation. Spending time in this particular setting afforded extracurricular interaction with a diverse set of peoples, including refugees and dignitaries from African countries who were visiting to learn about irrigation methods that they would apply upon return home to advance their societies. The impact of these experiences affirmed my passion for scientific discovery and provided clarity about how best to direct my career. This recognition led to enrollment in the graduate program in chemistry at Purdue University and receipt of a Ph.D. in this field.

During my dissertation work, I gained a deep appreciation for how the fundamentals of chemistry apply to complex macromolecules, particularly proteins. This fascination with understanding how the chemical attributes of diverse components direct protein interactions, establish three-dimensional structure, and impart catalytic function led me to pursue a postdoctoral position at Purdue in structural biology. The postdoc opportunity not only provided the ability to further develop as a scientist and gain expertise in a second technical area, but it created exposure to non-technical aspects of a faculty member's job, including supervising and mentoring junior trainees, project development, and proposal preparation. During both training periods, I participated in conferences, through which it also became evident that scientists operate both independently and as part of a network. The forum created opportunity to interact with other scientists over ideas and learn from more diverse perspectives. It also introduced me to individuals who have provided support and mentoring throughout my career path. Both formal training and informal mentoring contributed grately to my success, and I am a strong proponent of continuing to seek both throughout one's career.

I discovered, as many others did, that the interface between chemistry and biology represents a great opportunity for research and development. Upon completion of my formal training, good fortune prevailed, and I was presented with a unique opportunity to put my expertise to use in a new field. In 2004, I became a professor of Pharmaceutical Chemistry at the University of Kansas. In this position, I teach classes, train graduate students, conduct independent and collaborative research, and serve on committees that support various aspects of the university and scientific communities. The unique path I undertook has shaped my views and influenced how I engage my responsibilities as a faculty member.

The majority of my efforts as a professor are focused on research-related activities, and this is the part of the job that I enjoy most. My group focuses on understanding the behavior of proteins, particularly factors that affect stability and interactions, and we often collaborate with other experts to tackle more challenging problems. Through sustained efforts, my group and I have discovered and developed a unique metal-peptide chemistry, which has great potential to enable therapeutic and diagnostic healthcare applications. I founded a company,

Echogen, Inc., to commercialize the metal-peptide technology discovered in my academic research laboratory. Faculty do not often engage the business aspects of chemistry, but having served as a consultant to biotechnology and pharmaceutical companies provided a foundation for understanding the therapeutic development process and industry priorities. To support the success of the company, learning was enhanced substantively by interacting with a diverse of set of experts, including senior scientists, regulatory experts, attorneys, entrepreneurs, and consultants.

As I progress further in my career, I am increasingly involved with professional service. I have served (and continue to serve) in departmental committees. I established a relationship with Nairobi University in Kenya to enroll qualified candidates in our distance MS program. That experience has connected me to Kenyan scientists and afforded opportunity to travel to east Africa to better understand their aspirations and plans to advance education, pharmaceutical industry and clinical practice. I am also active in the American Chemical Society (ACS), currently serving as a member of the Committee on Science as the Chair of Public Policy and Communication.

Success Factors in an Academic Career

I have often been asked by students for advice on how to prepare for a faculty position, successfully advance in this career path, and sustain this career in the long term. Obviously, success in academia takes hard work, perseverance, and creativity. Above all, it is helpful to understand the constraints and opportunities in academia and plan accordingly.

Price of Admission

If you are interested in an academic career in a research university, you need a minimum requirement of a PhD in an appropriate field of study. Postdoctoral training has become a necessary component in preparation for an academic career in many fields. This two to four-year experience is an asset because it facilitates increased breadth of knowledge, enhances perspective on both scientific and professional issues that effect one's career, and builds credentials in forms that are commonly listed on one's Curriculum Vitae (CV). My predoctoral experience equipped me with analytical chemistry skills and expertise in protein structure and binding interactions, particularly using solution NMR. During this period, as is the goal, I became technically proficient at conducting a variety of experiments and performing data analysis. I learned how to read the literature and interpret my data with respect to the field and to present my findings to other scientists by drafting manuscripts and presenting posters. The postdoctoral experience provided the opportunity to design and plan research projects and identify gaps in knowledge that would be of benefit to fill. The structural biology group at Purdue was an excellent place to be a postdoc because the focused group of faculty, trainees and support staff were from diverse backgrounds and highly interactive, providing an incredibly rich intellectual resource to stimulate development.

Sometime during the academic training, you need to think ahead and investigate the career path that you would like to undertake. The key is to decipher your own career interests, understand your own strengths and weaknesses, and assess the job market for opportunities. If you desire to teach at a research university, you need to prepare an application packet, a CV and an independent research plan. These documents may take some time to develop, and you may want to revise and improve it as you learn more or notice gaps and holes. In my area, it was expected that a candidate would have at least 2-3 well-developed, fundable proposal ideas at the time of application, and that these ideas represent independent work as opposed to a continuation of a previous project. If the university job you are interested in has a substantial teaching component, you need to formulate a teaching philosophy that fits with your personality and is acceptable to the university. Most university faculty positions have a significant teaching component, which is the basis of your salary. Medical schools typically have lower teaching demands and as such provide less salary support, which faculty recover by winning grants. Because every school differs to some extent, my recommendation is to seek out a few strategic mentors in your field well in advance of the typical fall application cycle.

Before you apply for a job position, it is important to do your homework. Find out as much as you can about the university, the department, the existing faculty members, and the types of research that are being done. If you are offered an interview (or an actual job), you need to review the information again and gain further familiarity with the job and the stakeholders. Furthermore, it is never too early to investigate potential funding sources. If you have ideas about innovation and potential new products, you may want to explore possible mechanisms for commercial applications.

Successful Advancement

Most people start their academic career as an assistant professor. At most universities, formal review of performance is conducted in year 3, a process which often includes internal assessment, external review of research productivity by experts in the field, evaluation of teaching, and confirmation that a small amount of service has been performed, often to the department and scientific community. After 5-6 years the opportunity to apply for tenure and/or promotion to associate professor may be given, but this option is determined at the point of entry. The approach to tenure and promotion differs among universities, so it is advised to ask about the process. Future promotions to full professor, distinguished professor or an endowed chair may come in time as a result of continued productivity and level of success. Separately, one may be interested in administration and may pursue a path to become department chair, dean, provost, or president/chancellor of a university. In all cases, some general guidance can be given for successful advancement.

First, it is useful to have a career strategy. You need to understand your strengths and weaknesses, and the opportunities and challenges that you may face. You then need to set goals for yourself and map out a reasonable and achievable implementation plan. If your goal is to move quickly from assistant to full professor in 10 years, clearly determining all the things that you need to accomplish every year can enable that goal. Likewise, if you dream of becoming a university president, you need to seek administrative opportunities quickly. These two courses would require a very different strategic plan.

Secondly, be coachable. Your colleagues and friends may have good advice and they often are invested in your success. Seek input, listen carefully and respond appropriately. The decisions, of course, are yours, which is why it is helpful to get multiple opinions.

Thirdly, you need to engage your communities. Who are your communities? There can be a number of them. The professors in your institution represent one community. The people who are actively engaged in your research area constitute a highly significant group. The business stakeholders in your research area may be considered another. These may include companies who sell products in your area, testing labs that analyze those products, or venture capitalists interested in investing in your area of work. These are the communities to which you should pay attention because they influence your success, often in ways that are difficult to appreciate at the outset. If there is a formal organization, for example the ACS or a division within the society, or a trade group that represents your interest, it can be advantageous to selectively volunteer your time to support the success of that community. In general, it is helpful to find out who cares about your work and success and dedicate some effort to develop and nurture those relationships.

Fourthly, aim to make significant contributions to your field of study. We are scientists, and ultimately we are judged by the quality and the quantity of our work. Early on in my time at KU, a colleague offered some sage advice: Do good, solid, creative work, and make steady progress in your projects. It is commonly stated that perfection is the enemy of progress. Everyone wants to publish their work in a premier journal, but with regular reviews, it is crucial to show productivity. An apt analogy for this is that it takes many bricks to build a tower, so strive to make bricks early on and cement them together to build a tower over your career. This idea also applies to research funding. Great research is certainly what we all aspire to accomplish, but fundable research must be feasible for you to achieve at your institution within the time frame, and such practicality is necessary to bring in the resources to continue your research.

Fifthly, identify and seek out professional mentors. Most successful people have multiple mentors along the road to success. As you move along your career, you need to solicit and retain mentors. Mentors are not your equal; they are there to provide candid and objective advise. Although they do not fit the social category of friend, a good mentor has your long-term welfare in mind. They can provide useful perspectives, open doors when needed, and enable thoughtful decision making. In other words, they can alert you to the unknown unknowns.

Sustainability

From the ongoing account, it is clear that career development takes a lot of work. How do you sustain yourself for many years without fatigue or burn-out? It is important at the outset to determine what you want from your career. If this career path provides satisfaction, you are more likely to endure the difficulties that are encountered along the way. An academic career is not for the faint-hearted. It helps to set clear priorities and boundaries. There is always work to be done. If you can prioritize the various tasks at hand, you ensure that the most important things get done. Most of us have other obligations beyond the workplace that place demands on our time and energy. Think comprehensively about your responsibilities and define the hierarchy of your priorities, including carving out time to recharge and care for yourself. Establishing a set schedule and finding ways to protect your time from distraction is important when you have numerous demands. Email, for example, can consume your entire day. It may keep you busy but also may result in very low productivity. The academic's career is many decades long, and it is helpful to treat it like a marathon and pace yourself.

Some colleagues will not agree with me on this point, but I have become a firm believer that it necessary to take time off periodically to recover your energy and refresh your mind. The manifestation of time off may not look at all the same from one person to another. Everyone has limits and only you know yours, so determine what you need to do for yourself and accept that other opinions may differ. In the end productivity is what gets evaluated. I also think it is important to be optimistic. A positive attitude is infectious. If you look good and feel good, the people around you also tend to feel good. If someone criticizes you, try not to take offense. Separate emotion from fact; listen for the message. If the criticism is justified, embrace it and put it to good use.

Finally, in today's world, the only thing constant is change. Change is not necessarily bad; it can actually be useful. The important thing is to adapt, evolve, and enjoy!

Conclusions

As a faculty member, you will teach, conduct research, and train students. You may advance your career in a variety of ways, for example through research or choosing to get into university administration. At the same time, you can engage in business activities by getting involved as an industry consultant or even starting your own business. In any case, it is useful for you to make individual and institutional interconnection with partners outside the university, as these relationships are critical to effective translation of and, increasingly, funding for research. In a way, we can think of a faculty position as equivalent to the owner of a small business. A business needs products that can be sold; it needs to market its products, and it needs to manage its finances. The same is true for a successful faculty member. The learning offered in these pages has helped me succeed in and enjoy my academic career, and I hope the examples and advice provided here prove useful to you in considering the academic career path.

References

1. Gabrys, B. J.; Langdale, J. A. *How to Succeed as a Scientist: From Postdoc to Professor*; Cambridge University Press: Cambridge, 2011.

2. Coley, S. M.; Scheinberg, C. A. *Proposal Writing: Effective Grantmanship*; Sage Publications: Thousand Oaks, CA, 2008.
3. Collins, L. H.; Chrisler, J. C.; Quina, K. *Career Strategies for Women in Academe*; Sage Publications: Thousand Oaks, CA, 1998.

Chapter 4

Industrial Careers: Obtaining a Job and Succeeding in Industry

Joel I. Shulman*

Department of Chemistry, University of Cincinnati, P.O. Box 210172,
Cincinnati, Ohio 45221, USA
(The Procter & Gamble Company, Retired)
*E-mail: joel.shulman@uc.edu

Obtaining a job in industry and then succeeding in it takes a somewhat different skill set than succeeding in academe. Industrial interviewers probe for professional skills such as teamwork, nontechnical communication skills, leadership potential, and risk taking, in addition to the more obvious technical and problem-solving skills. Information on these skills and other "key performance factors" is developed using a technique known as behavioral interviewing. Once you have obtained an industrial job, you can get off to a fast start and begin a successful career by learning the culture of the company, networking, communicating well, and, of course, working hard and smart. This paper will outline what companies are looking for in their future employees and will suggest strategies for embarking on a successful career.

Approximately 54% of PhD chemists in the United States work in industry (*1*). For bachelor-level chemists, the percentage is even higher. Yet students rarely receive training in how to interview for industrial jobs and even more rarely learn what is valued by industry. The recent summary report of an American Chemical Society Presidential Commission, *Advancing Graduate Education in the Chemical Sciences* (*2*), presented as its first conclusion "Current educational opportunities for graduate students…do not provide sufficient preparation for their careers after graduate school." The report cited the need to enhance students' ability in several areas that are highly valued by industry:

- Communicate complex topics to both technical and nontechnical audiences, and to effectively influence decisions;
- Learn new science and technology outside prior academic training;
- Collaborate on global teams and/or with global partners and clients;
- Effectively define, drive, and manage technical work toward a practical, significant result.

Candidates for jobs in academe are evaluated almost exclusively in two areas: their demonstrated ability and future potential to perform outstanding research and/or their ability to teach. To obtain a job in industry, being able to conduct meaningful research is a *sine qua non*; it is a necessary, but not by itself a sufficient, ability that must be demonstrated to get hired. So what must an applicant know and what are the critical skills that an applicant must demonstrate in order to be a strong candidate for industry?

Résumé *vs.* CV

Any job search begins with a résumé or *curriculum vitae* (CV). The purpose of either is to get you an interview; both are marketing tools, where you are the item being marketed. However, résumés and CVs can be very different. *Curriculum vitae* is Latin for "course of one's life." It is an account of what you've been up to in your professional life—in detail. A résumé, while also describing your professional life, is by definition short and focuses the reader on those aspects of a person's background that are directly relevant to a particular job. Academic employers expect to see a CV as part of an application package, while industrial employers look for a concise and well-crafted résumé. Figure 1 compares and contrasts résumés and CVs.

It should be noted that most applications for industrial jobs occur through the internet and applications/résumés are often searched via software looking for keywords in context. Thus, the résumé should contain keywords targeted to specific positions of interest. Similarly, employers routinely use social media (*e.g.*, LinkedIn, ACSNetwork) and Google searches to learn about applicants, so it is important to keep your internet presence up to date and a positive reflection of you.

Successful Interviewing for Industry

The whole purpose of interviewing candidates for a job in industry is to determine if the person:

- *Can* do the job (has the technical competence/mastery, additional professional skills, and scientific fit to be successful)
- *Will* do the job (has the desire, motivation, and work ethic to be successful)
- *Fits* into the company's culture and organizational style

Résumé vs. Curriculum Vitae

Résumé	CV
• Quick read of skills and accomplishments	• Complete listing of activities and accomplishments
• Get an interview, not a job	• Get an interview, not a job
• Guide during the interview	• Guide during the interview
• Must be clear, concise, and accurate	• Must be clear and accurate, not necessarily concise
• Two page maximum (plus page for publications)	• Continues to grow throughout your career
• Can vary according to the job sought and replace information over time	• Does not eliminate information over time, just adds to what is there
• Should contain key words that will be found during a computer search	

Figure 1. Comparison of a résumé and a CV

It is imperative that candidates for industrial jobs understand that their research experience and accomplishments alone will not land them a job. They must demonstrate professional skills that predict success in the job. These skills can be termed "Key Performance Factors" (some employers call them "Critical Success Factors") and are probed by an employer during the interview process. Every company has its own definition of Key Performance Factors, which are designed to be consistent and objective standards against which all candidates can be measured. Ideally, the company has assessed how its top-performing employees measure up in each of these Factors. Then, all candidates are formally rated on each Factor. What are these Key Performance Factors and how can a company assess them in each candidate?

Key Performance Factors

A typical set of Key Performance Factors could be the following (Technical Mastery is the *sine qua non*; the remaining Factors are presented in alphabetical order):

TECHNICAL MASTERY
- A solid understanding of your field
    - Experimental design
    - Laboratory skills
    - Data interpretation and modeling

- Professional independence
- Ability to convert technical skills into practical application

CAPACITY
- Demonstrate ability and motivation to work hard and smart
- Pursue learning, keep up with emerging trends
- Seek out and eliminate wasted effort
- Set appropriate priorities
- Demonstrate strong initiative and follow-through

COLLABORATION [Understand that industry solves problems in teams]
- Ability to work in a team environment
- Effective interactions with a diverse group of colleagues
  - Diverse in personal background
  - Diverse in technical training
- High personal standards and integrity
- Solid oral and written communications skills

INNOVATION
- Go beyond accepted ideas
- Make connections, reapply knowledge and approaches
- Convert new ideas into workable experiments or solutions
- Identify better ways to do things

LEADERSHIP [Envision, Engage, Energize, Enable, and Execute]
- Recognize opportunities and show vision
- Recognize issues and develop strategies to meet them
- Set direction and capture commitment (be a champion)
- Identify and use resources effectively
- Resolve conflicts
- Get the job done

RISK TAKING [Intelligent risk taking, not taking physically dangerous chances]
- Set specific, stretching goals
- Display a sense of urgency in achieving goals
- Take informed risk: Be assertive in making decisions in the face of uncertainty
- Understand the 80:20 rule (modification of Pareto's Principle) (3) as it relates to getting the job done
  - You can complete 80% of a project in 20% of the time that it will take to dot all the "i's" and cross all the "t's" to answer every last question.

SOLUTIONS
- Identify/define problems and evaluate alternatives

- ○ Understand that it is imperative to define a problem well: solution of a poorly defined problem will not help the business goals of a company
- Sort through complex data and draw appropriate conclusions
- Integrate data and intuition from a range of sources
- Implement next steps

Behavioral Interviewing: An Approach to Assessing Key Performance Factors

Obtaining information from a candidate about Key Performance Factors is usually accomplished using a deceptively simple process call "behavioral interviewing." This is an approach based on the theory that the most accurate predictor of a person's future performance is past performance in a similar situation. In this approach, the employer has established a correlation between behavior relative to Key Performance Factors and success on the job. The employer then develops questions to probe a candidate's behavior in certain situations. Ideally, interviewers are trained to evaluate answers so that there is consistency across the organization among interviewers.

Typical behavioral-based questions are rather transparent:

- Tell me about a difficult problem you faced in your research and how you solved it. [Solutions]
- Describe for me a job experience when you had to serve as the leader in order to accomplish a goal. [Leadership]
- When working on a team, what role do you usually take? Why? [Collaboration]
- Give me an example of a time when you have had to handle a variety of assignments. Describe the results. [Capacity]
- Describe the most creative work-related project you have carried out. [Innovation]
- Describe a time when you set your sights too high (or too low). [Risk Taking]

An excellent way to frame your answers to behavioral-based questions is to "Build a **CAR**." That is, frame your answers around:

- Context
- Action
- Results

What was the situation that led to the example you are going to use and why was it important? What action did YOU take? And, what were the results?

In addition to behavioral-based questions, many interviewers like to ask you to elaborate on your strengths and weaknesses. The former is an excellent opportunity for you to tell the interviewer why she should hire you. For the latter, you need to have a couple of examples of areas in which you would like to improve yourself—not "fatal flaws," but areas you have identified and developed a plan to overcome.

Before any interview, it is a good idea to practice answering sample questions *out loud*, either with a friend or by yourself. Just thinking about how you will answer a question is not sufficient; articulating your answer is more difficult and takes practice. There are many Web sites that can provide you with the types of behavioral questions you might encounter, for example the one by Denham (*4*).

To enhance your candidacy for an industrial job, doing informational and networking interviews (meetings with an employee in a job similar to the one you are seeking) before your actual interview will help define what might be sought at a firm and what it is like working in industry.

Figure 2 summarizes the steps you can take to prepare yourself for an industrial interview.

Preparing for Interviews

- Think about what you have done in your personal life—graduate school and otherwise.
- Jot down accomplishments and the skills that led to them.
- Compartmentalize accomplishments according to Key Performance Factors (KPF).
- Try to identify accomplishments/examples for each KPF.
- Think in depth about your strengths and weaknesses.
- Do some practice interviewing.
- If possible, learn about the requirements of the job.
- Remember CAR: Context, Action, Results.

Figure 2. Steps to take in preparation for interview

Getting Off to a Fast Start in Industry

The first few months at an industrial job will set the tone for your career. Getting off to a fast start will not guarantee success with a specific employer, which depends on many factors, but will improve your chances of succeeding in the long term. You should pay attention to four items in particular:

- Learning the company culture
- Becoming adept at networking
- Communicating effectively
- Working hard and working smart

Learning the Company Culture

A company's culture is its attitudes, norms, and accepted values. Culture can be seen in a company's approach to research (Is it cutting edge, derivative, or "me-too"? Is publishing encouraged? Does it lead to important patents?); its economic situation (Does the company spend money to save time or spend time to save money?); and personal interactions (Are they formal or informal? Do employees socialize off the job? Is face-time on the job important or are you trusted to set your own schedule as long as the job gets done?). Figure 3 gives examples of the kinds of questions you can ask to help discover a company's culture.

Discovering a Company's Culture

Examples of questions to ask:
- What's important at work?
- What things are mandatory? Optional?
- What skills and characteristics are valued?
- How flexible is the length of a typical work day?
- What are the criteria used for promotion?
- How is professional development encouraged?
- How do I get feedback on my performance?

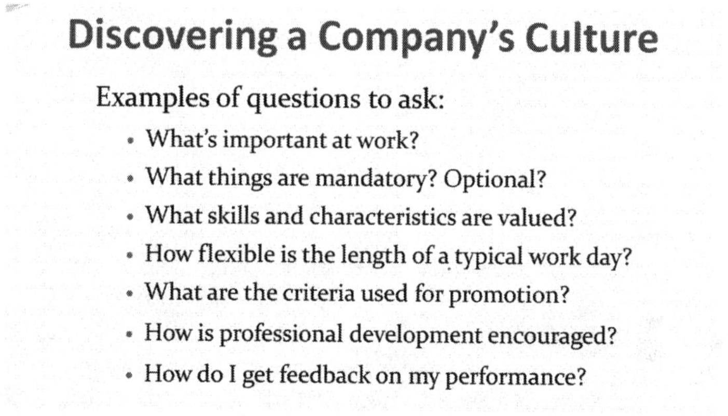

Figure 3. Questions that can be used to discover a company's culture

Understanding company culture and beginning to fit into it is important. Fitting in will go a long way toward guaranteeing that you will be respected and your ideas will be heeded. This will facilitate having growth opportunities within the company.

Becoming Adept at Networking

Networking is making connections with others to exchange Information, Ideas, and Introductions. It has been defined as "the development of a mutually supportive, interactive, ever-widening circle of those with similar objectives, abilities, and views" (5). Networking is an important skill to develop before you look for a job and to continue once you have started a job. Indeed, it should continue throughout your career.

Networking is an art, not a science. It must involve giving as well as receiving. It should begin in your own company, extending across organizational lines. This will enable your understanding of company culture to grow. But beyond that, it will help you understand other opportunities that may exist in the company: what others do on a daily basis, who they interact with, what types of things they need to know. However, networking should not stop with your coworkers. Networking with professionals outside of your company is an effective means for personal development. This can occur at local or national meetings of professional societies; during interactions with suppliers or vendors; or as a volunteer with a charitable organization.

A subset of networking is mentoring. A mentor is a "loyal advisor" who knows the ropes of the company and is willing to spend some time with you to facilitate your growth within the company. Identifying a mentor within your company is another way to get off to an effective start in your career. In fact, having more than one mentor can be particularly valuable. It has been said that "All professionals (and younger ones in particular) should try to develop three types of mentors. One should be a supervisor, team leader, or experienced coworker who can provide technical advice. Another should be a coworker skilled in the business areas of your company...The third type of mentor should be someone outside your company who is knowledgeable about the industry or technical field in which you work" (6).

Communicating Effectively

The ability to communicate effectively is vital to success in any career. Within industry, this includes oral and written communications with your boss and with coworkers, and importantly with people outside of your field—including nontechnical people. For example, it is not uncommon in industry for chemists to be called on to present their work to managers on the business side of the company. In this case, effective communication with nontechnical people may determine the fate of a project.

Communicating with your boss may be the most important communication job you have when you start an industrial job. As noted above, your supervisor should serve as a mentor for you, helping you learn the ropes of the company: what is important, what is expected, how to operate day-to-day. To communicate with your boss most effectively, you need to develop some basic information about her preferred style of interaction. Does she micromanage; act as a hands-off manager, seeking information only on an as-needed basis; or something in between? Does she prefer informal meetings with mainly oral interactions or does she like to receive all important information in writing? What is her balance between being 100% bottom-line oriented and being a very people-oriented supervisor?

A major part of your job is to make your boss look good! (In fact, one of the best ways to get promoted is to help your boss get promoted in front of you.) Remember that she has a boss, as well. Therefore, you need to provide her with data/information that she can use to communicate up the management line, keeping her informed about all important aspects of your work. You will probably have scheduled (weekly, biweekly, etc.) meetings with your boss. You should prepare carefully for these, having a written agenda and thinking about what you want to accomplish at each meeting. When you have a problem, technical or otherwise, you don't want to walk into your boss's office and toss the problem on her desk. Rather, you want to provide possible solutions that you have thought about and ask for her perspective, not for "marching orders" that you should follow.

It is difficult to over communicate in industry. This is especially true of written communications. Timely production and dissemination of reports, especially when they are not "required," will reflect positively on you. For example, providing a written summary of an important meeting to all participants will be a valuable document for everyone and will demonstrate your proactivity. Of course, all documents must be carefully proofread and be error free.

Good companies set direction based on good proposals, and the best companies run on proposals developed from the bottom up. Being able to convince your management of something, even something relatively trivial (*e.g.*, purchasing a new piece of equipment), often demands a written proposal. A valuable model for presenting any argument, particular a written proposal, goes by the acronym SOPPADA (Figure 4) (*7*).

"SOPPADA": An Effective Template to Convince in Writing

- Subject of the communication
- Objective of the communication
- Past (Background)
- Proposal
- Advantages of proposal
- Disadvantages of proposal
- Action steps

Figure 4. SOPPADA - A good template for a written document

The heart of a SOPPADA communication is the Proposal itself. But the **Advantages**, **Disadvantages**, and **Action Steps** are vital to influencing the audience you are trying to convince. For the Advantages section, you need to know the "hot buttons" of your audience. For example, if you are trying to obtain buy-in to spend money for a new instrument and your target audience is both budget and schedule oriented, emphasize the time savings that will accrue to projects with the new purchase. Similarly, describe the strongest possible Disadvantage and present an argument directed to muting this Disadvantage. Thus, in the above situation, cite the fact that the purchase may cause the budget to take a short-term hit but in the long term it will save money by saving time. Finally, the Action Steps should not simply be the approval of the Proposal. Rather, they should include the very next step in implementing the Proposal, *e.g.*, "attached for your approval is a purchase order for this instrument." Following the SOPPADA style allows you to make a cogent and succinct argument for any proposal.

Working Hard and Working Smart

The majority of chemical professionals—certainly those with a PhD—will be hired by industry as "exempt" employees. This means they are exempt from the wage and hour requirements of the Fair Labor Standards Act. Exempt employees do not get paid overtime. They are expected to do whatever it takes to get the job done, even when, as often is the case, that requires working more than a 40-hour week and/or working at home in the evening or on weekends. The goal of every employee should be to exceed expectations—work hard to go above and beyond what the "average" employee would produce.

Setting personal objectives for your work is one approach that can help you exceed expectations. A useful paradigm for setting such objectives is to make them SMART: Specific, Measureable, Achievable, Realistic, and Time-based. With a particular goal in mind, answering the questions in Table 1 can help determine if your objectives reflect the SMART criteria (8).

Table 1. SMART objectives

SPECIFIC	_MEASURABLE_	_ACHIEVABLE_	_REALISTIC_	_TIME-BASED_
• Who is the target population? • What will be accomplished?	• Is the objective quantifiable? • Can it be measured? • How much change is expected?	• Can the objective be accomplished in the proposed time frame with the available resources and support?	• Does the objective address the goal? • Will the objective have an impact on the goal?	• Does the objective propose a timeline when the objective will be met?

Let us say, for example, that you have been assigned a project to develop a new catalyst for a particular purpose. You might set as your first objective to review the pertinent literature, both in-company documents and external publications, and within six weeks produce a report for your management that proposes an approach to the problem. Such an objective is Specific (the accomplishment is a report for your management); Measureable (the report is the product of your effort); Achievable (your company library has access to both internal documents and external literature); Realistic (this is a standard approach to developing a new project); and Time-based (a report in six weeks).

Most companies have a formal process by which to assess performance—whether working hard and working smart is leading to success. While informal performance reviews can occur anytime and anywhere, and provide immediate feedback, formal reviews are usually regularly scheduled with a standardized structure. Generally, the structure consists of several steps:

- Data gathering, including a review of accomplishments and contributions as well as strengths and areas for improvement.
- A discussion of the data with management, often your supervisor and his boss.
- Agreement on the value of your efforts.
- Discussion of areas for personal development as well as your future objectives.
- Agreement on who (you and/or your managers) will do what in terms of continuing your personal development and helping you achieve the career you desire.
- Written documentation of the discussions and action plans.

Formal performance reviews are often annual occurrences. Because some of the information needed to document accomplishments and contributions may therefore be a year old, it is helpful to keep a contemporaneous journal of your major activities on an ongoing basis. This journal does not have to be detailed, but will serve to remind you of your successes and what led to them when performance-review time comes up.

The impact of performance reviews on your career can be large. Your performance and contributions are compared with those in the company at an equivalent organizational level, so the data produced in a review will impact your salary increases and opportunities for promotion (Figure 5). Salary growth is relatively rapid for ~10 years with a company, after which it levels off in inflation-adjusted terms, with growth occurring more slowly. Most companies link pay with performance: If you raise your performance, you can get a higher salary and extend the growth phase for several additional years. This is your reward for excellence—for exceeding expectations—as seen in the chart below. On the other hand, failing to meet expectations on a continuing basis can lead to termination.

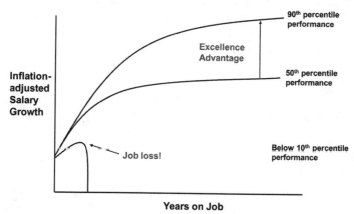

Figure 5. Salary growth as a function of years on job and performance. Courtesy of the American Chemical Society.

Summary

To obtain a job and be successful in industry, you need to:

- Understand and prepare for the interview process and recognize how it differs from finding an academic job.
- Get off to a fast start by understanding the company, developing networks and mentors, communicating effectively, and working both hard and smart.

References

1. National Science Foundation, National Center for Science and Engineering Statistics, 2012. *Characteristics of Doctoral Scientists and Engineers in the United States: 2008*; Detailed Statistical Tables NSF 13-302; Arlington, VA. Available at http://www.nsf.gov/statistics/nsf13302/.
2. *Advancing Graduate Education in the Chemical Sciences, Summary Report of an ACS Presidential Commission*, 2012. http://www.acs.org/content/dam/acsorg/about/governance/acs-commission-on-graduate-education-summary-report.pdf.
3. Koch, R. *The 80/20 Principle: The Secret to Achieving More with Less*; Doubleday: New York, 2008.
4. Denham, T. 50 behavioral-based interview questions you might be asked. http://blog.timesunion.com/careers/50-behavioral-based-interview-questions-you-might-be-asked/1538/.
5. Sinderman, C. J. *The Joy of Science*; Plenum: New York, 1985; Chapter 4.
6. Borchardt, J. K. *Career Management for Scientists and Engineers*; Oxford University Press: New York, 2000; Chapter 8.
7. Brucker, R. W. SOPPADA: A handy way to present ideas, proposals, suggestions, and plans. http://www.rogerbrucker.com/soppada.pdf.
8. Writing SMART, Short-term Outcome Objectives. http://www.azed.gov/century-learning-centers/files/2011/06/16.-writing-smart-short-term-outcome-objectives.pdf.

Chapter 5

Career Opportunities in the Government Sector

Kristin M. Omberg*

Center for Integrated Nanotechnologies, Materials Physics & Applications Division, Los Alamos National Laboratory, PO Box 1663, Los Alamos, New Mexico 87545
*E-mail: komberg@lanl.gov

A 2010 survey of American Chemical Society (ACS) members indicated that nearly 10 percent were employed by federal, state or local government entities. ACS members work for a wide range of different government and government-affiliated organizations, including the United States Congress and the Congressional Research Service; the federal agencies; the National Academies of Science and Engineering, Institute of Medicine and National Research Council; and state and local government bodies including those concerned with environmental and public health. These ACS members live all over the United States, and perform a wide range of job functions, including policy analysis and program management; research and development; quality assurance and quality control functions; and sample analysis activities. This article presents a brief overview of opportunities available in the government sector, with links to additional information for those interested in more detail.

Introduction

A 2010 survey of American Chemical Society (ACS) members indicated that nearly 10 percent were employed by federal, state or local government entities. ACS members work for a wide range of different government organizations, including the United States Congress and the Congressional Research Service; the federal agencies; the National Academies of Science and Engineering, Institute of Medicine and National Research Council; and state and local government bodies including those concerned with environmental and public health. These ACS members live all over the United States, and perform a wide range of job functions, including policy analysis and program management; research and development; quality assurance and quality control functions; and sample analysis activities. Salaries and benefits are typically competitive with the broader marketplace, and foreign nationals may be eligible to apply, particularly for entry-level positions. This article presents a brief overview of opportunities available in the government sector, with links to additional information for those interested in more detail.

Career Opportunities in the United States Congress

Science, technology, engineering, and math (STEM) professionals can add tremendous value to the legislative process by serving as interns, fellows, or staffers in Congressional offices. There are even a few STEM professionals who have been elected to the United States (US) Senate and House of Representatives (though, in the opinion of the author, not enough!). Congressional staff may support a personal office (the office of an individual Senator or Representative) or a Committee (for example, the US Senate Committee on Commerce, Science & Transportation or the House of Representatives Committee on Science, Space & Technology). Staff positions are exciting and fast paced, changing sometimes daily as different legislation is introduced for consideration. They require general STEM knowledge and training, strong communications skills, flexibility, and the ability to multi-task.

Personal office staff have the opportunity to interact regularly with both their Member (the Senator or Representative whom they serve) and the people whom their Member represents. One of the most important duties of a Member is to listen, and respond, to his or her constituents; all of the staffers in a personal office support the Member in this vital work. Personal office staff usually work on multiple issues of interest to the Member's District or State simultaneously, and must be able to advise the Member on policy alternatives that are sound and supportable, but also best represent a wide range of constituent views. Constituent concerns may be scientific in nature, but they may be economic, religious, ethical, or driven by any number of other factors. STEM professionals in personal offices have a unique opportunity to view current issues—including science policy issues—through a wide range of perspectives, and advise their Member on the best courses of action for all.

Committee staffers typically work in a more circumscribed set of areas, as Committees specialize in particular types of legislation (whereas individual Members attend to all legislation brought to the floor). For example, the House Committee on Science, Space & Technology and the Senate Committee on Commerce, Science & Transportation regularly authorize the programs executed by the National Science Foundation (NSF) and the National Aviation & Space Administration (NASA), as well as other Agencies' scientific activities. Authorization bills define the priority areas upon which the agencies will focus in upcoming years, so staffers on these oversight committees specialize in staying abreast of specific scientific programs at these agencies and their progress.

One of the easiest ways to find out about opportunities for STEM professionals in Congress is to contact your own Congressional Delegation. Senators and Representatives are often looking for volunteers at the State or District and National levels, and these (usually unpaid) positions (even short-term ones) provide great conduits to paying jobs. Job postings for both paid and unpaid positions can also be found at www.senate.gov and www.house.gov.

STEM professionals with advanced degrees are eligible to participate in a number of paid Science Fellowship programs. Almost 50 STEM professionals each year work on Capitol Hill through programs administered by the American Association for the Advancement of Science (AAAS; see fellowships.aaas.org for more information). The American Chemical Society (ACS) sponsors two Congressional Fellowships under the AAAS umbrella; more information on these positions is available at www.acs.org/content/acs/en/policy.html. These programs provide a fantastic opportunity for STEM professionals to participate in the policy process and determine whether this is their long term career path.

Career Opportunities in the Congressional Research Service

The Congressional Research Service (CRS) is a branch of the Library of Congress that provides in-house research reports on policy topics under consideration in the current Congress. It employs researchers in a wide range of areas, including STEM professionals.

Recent CRS reports relevant to science policy have included "Science, Technology, Engineering, and Math (STEM) Education: A Primer," published August 1, 2012; "Rare Earth Elements in National Defense: Background, Oversight Issues, and Options for Congress," published March 31, 2011; and "Oil Spills in U.S. Coastal Waters: Background, Governance, and Issues for Congress," published April 30, 2010. Although the CRS does not make these reports publicly available on its website, many reports can be found at second-party sites such as www.opencrs.com or www.fas.org/sgp/crs/row/. Employment opportunities at CRS can be found online at its official site, www.loc.gov/crsinfo.

Career Opportunities at the Federal Agencies

There are many positions in federal government agencies staffed by STEM professionals. The US Departments of Agriculture, Defense, Energy, and Homeland Security; the US Environmental Protection Agency (EPA); NASA; the National Institutes of Health (NIH); the National Institute of Standards & Technology (NIST); NSF; the Patent & Trademark Office; the Chemical Safety Board; and the intelligence agencies (among others) all employ individuals with STEM degrees.

STEM opportunities at the federal agencies fall into two general categories: policy & program management, and research & development. Policy staffers analyze and implement legislation and policy. For example, EPA has responsibility for implementing the Clean Air Act; the Department of Homeland Security (DHS) oversees many chemical safety programs; and the Department of the

Interior oversees the extraction of natural resources and minerals from Indian and non-Indian federal lands. All these issues require the insights of STEM professionals for effective policy implementation.

Program management staffers oversee the execution of a variety of federally sponsored programs, including research and development activities. At the Defense Threat Reduction Agency, which is part of the Department of Defense (DOD), STEM professionals oversee research portfolios focused on, for example, the development of new materials, sensors, and therapeutics to support the warfighter. At the Department of Energy (DOE), STEM professionals oversee research portfolios relevant to energy and defense issues, such as the development of sustainable energy technologies. At the NIH, STEM professionals oversee research in many areas related to developing a greater understanding of all things health. And at the Federal Bureau of Printing & Engraving (part of the Department of the Treasury), among other things, STEM professionals implement and manage quality assurance and control programs to determine the acceptability of raw ingredients and materials required to (literally) make money. STEM professionals bring critical perspective to these positions because they understand the associated scientific challenges, and can determine how these challenges may impact the agencies' priorities.

Postings for program & policy management positions at the federal agencies can be found at www.usajobs.gov, the federal employment website. In addition, specific agencies, such as NSF, offer temporary or rotational opportunities that are often publicized on their websites. The AAAS also administers a one- to two-year fellowship program that places STEM professionals with advanced degrees in policy & program management positions at a variety of federal agencies, including the EPA, DOD, DHS, NSF, the Department of State and the US Agency for International Development (see fellowships.aaas.org for more information). And the U.S. Office of Personnel Management sponsors the Presidential Management Fellows program. Its new "STEM track" places recent recipients of advanced degrees in two-year rotational assignments designed to encourage a new generation of federal leaders (http://www.pmf.gov/). All these positions provide exceptional opportunities for STEM professionals, many of whom come from more traditional research & development backgrounds, to get a feel for what it is like to have a career in program & policy management.

Another, slightly different opportunity for STEM professionals is the US Patent & Trademark Office (USPTO). The USPTO employs about 2500 STEM professionals as Patent Examiners, who determine whether an invention is patentable. Patent Examiners specialize in particular areas, and there is a great need for trained STEM professionals to participate in this important process.

A number of federal agencies also have in-house research & development capabilities targeted toward their particular mission. For example, DOD employs researchers who work on issues critical to sustaining US defense capabilities; NIST (which is part of the US Department of Commerce) employs researchers who work on improving standards to facilitate commerce. The agencies' research laboratories fall into two broad categories: government-owned, government-operated (GOGO), and government-owned, contractor-operated (GOCO). GOGO employees are direct employees of the federal government.

DOD, EPA, NASA, and NIST are among the federal agencies with GOGO research & development components. Job opportunities at GOGO laboratories are posted at www.usajobs.gov.

GOCO laboratories are managed by a contractor, but perform research & development largely in support of federal agencies. GOCO laboratories may be referred to as Federally Funded Research & Development Centers (FFRDCs), which are research & development organizations that receive more than 70 percent of their funding from the federal government (usually one agency). DOE's National Laboratories are a prominent example of a GOCO system; they are principally funded by DOE or the National Nuclear Security Administration, but also perform work for other national security customers (including DHS and DOD). The Jet Propulsion Laboratory at Caltech, sponsored by NASA, and Lincoln Laboratory at the Massachusetts Institute of Technology, sponsored by multiple agencies, are also well known GOCO entities. Opportunities at GOCO laboratories are usually posted at each individual laboratory's website; for example, opportunities at Los Alamos National Laboratory can be found at www.lanl.gov.

Many federal agencies, and GOGO and GOCO laboratories, offer internships and fellowships for STEM professionals. Postings can be found at www.usajobs.gov (for GOGO positions), or an individual GOCO laboratory's website. Some of these opportunities are paid; all provide a wonderful opportunity for an individual to develop a network inside the agency or laboratory.

Career Opportunities at the National Academies of Science & Engineering, Institute of Medicine, and National Research Council

The National Academies of Science & Engineering, Institute of Medicine, and National Research Council are private, non-profit entities chartered by Congress to provide "independent, objective, and non-partisan advice with high standards of scientific and technical quality" on matters of national and global importance. While the Academies are not part of the federal government, they are commissioned by federal agencies, Congress and private entities to perform studies on critical issues in science, engineering and medicine that impact the government's effectiveness or ability to execute its mission. The studies are performed by panels of volunteers, typically highly esteemed STEM professionals with expertise in particular areas relevant to the study. The studies are facilitated by in-house staffers, many of whom are also STEM professionals. The staffers identify and organize expert panels, and facilitate their research, meetings, report production, and peer review. The National Academies Press makes most of its reports and committee proceedings available to the public; many are available for free download at www.nap.edu. Job postings are listed at www.nationalacademies.org/humanresources.

The National Academies also sponsor internships for STEM professionals interested in investigating public policy opportunities. The Christine Mirzayan Science & Technology Policy Graduate Fellowship Program is a 12-week program that allows graduate or professional school students and recent graduates to engage in studies throughout the Academies, and to develop skills in the science policy area. The Lloyd V. Berkner Space Policy Internships provide "promising undergraduate and graduate students with the opportunity to work in the area of civil space research policy" in concert with the National Academy of Science's Space Studies Board. The Public Interfaces of the Life Sciences Internship Program for undergraduate and graduate students "seeks to provide a real-world experience in science communication, science journalism, and science policy." Information on these programs, as well as other internship opportunities, is available at www.nationalacademies.org.

Career Opportunities in State and Local Governments

A variety of opportunities are available for STEM professionals in state and local governments, in many of the same areas as those at the federal level. Many states and major metropolitan areas have Departments that oversee environmental concerns, such as a Department of Environmental Health or Management. STEM professionals in these organizations may implement state or local policies (for example, policies dealing with air pollution) or may work in analytical laboratories (for example, analyzing air samples for contaminants). State and local Health Departments frequently employ STEM professionals in policy & program management, as well as in analytical laboratories. And STEM professionals can play important roles in state Departments of Education, as much of the education policy in the US is set at the state (not national) level. Postings for state or local positions can be found at individual entities' websites.

Salary and Benefits

Salaries for government employees are usually easy to benchmark, as there is much public information available. Congressional salaries are set by individual offices, and range between $38,000 (entry level) to a maximum of $172,000. Members of Congress earn upwards of $174,000.

Federal salaries and benefits are set by the Office of Personnel Management (www.opm.gov). Federal employees may be General Schedule (GS) employees or part of the Senior Executive Service (SES; a higher level of career federal employees with advanced skills). Entry level positions for new PhD scientists are usually leveled at GS-12; in 2013, the salary for a GS-12 was about $75,000 per year. The maximum salary for an SES is about $180,000 per year. More complete salary schedules can be found at www.opm.gov/policy-data-oversight/pay-leave. Federal benefit packages are often extremely attractive, and may include recruitment, relocation, and retention incentives, including student loan repayment. More information on benefits can also be found at www.opm.gov/policy-data-oversight/pay-leave.

Salaries and benefits at GOCO and FFRDC laboratories are set by the management contractors. They are generally competitive with federal and market salaries for comparable jobs. Information on GOCO and FFRDC salaries and benefits can be found at the individual laboratories' websites. State and local salaries and benefits are also set by the individual states and local entities.

A less obvious benefit of government employment is the ability to live in a variety of different physical locations. While it is true that many government opportunities for STEM professionals are centered in the greater Washington, DC area, government positions are located all over the United States. The EPA maintains facilities in Cincinnati, Ohio and Research Triangle Park, North Carolina. NIST has laboratories in Germantown, Maryland and Boulder, Colorado. DOE and the National Nuclear Security Administration maintain laboratories and facilities in more than a dozen states, including California, Idaho, Illinois and Tennessee. And the Agricultural Research Service, the research arm of the US Department of Agriculture, has laboratories in almost every state in the US. Opportunities are also available for federal employees to move between agencies and locations, so an entry level position in one location or agency may well provide an avenue for a transfer to another.

Conclusions

Many opportunities are available for ACS members interested in pursuing a career in the government sector. STEM professionals may find employment in a wide range of government or government-affiliated organizations. Positions may be located anywhere in the US, and may involve policy analysis & program management; research & development; quality assurance and quality control functions; or sample analysis activities. Salary and benefit packages are usually competitive with the general marketplace, and employment in one government entity or location may open up opportunities for employment in another.

Although many government positions require US citizenship, it is not a requirement for all opportunities. Entry level positions, in particular, may be open to foreign nationals. Citizenship requirements are usually specified in job postings.

A final opportunity that has not been covered elsewhere in this article is ACS's Science Policy Fellowship. ACS sponsors one Fellow annually to work in ACS's Office of Public Affairs (OPA). OPA handles ACS's government relations work, including interactions with the federal agencies, Congress, and other scientific societies. This Fellowship allows ACS members with advanced degrees to work with experienced ACS staff to provide information to policy-makers on the role of science in public policy; present specific policy recommendations on issues affecting ACS members; and inspire ACS members to engage in the policy process. OPA staffers handle a range of issues of interest to ACS members, including federal funding for scientific research; STEM education; and environmental policy.

Chapter 6

From the Lab to a Career in Business Development – An Unexpected Career Transition

Mark D. Frishberg[*]

Custom Organic Synthesis Consultant and Vice President, Business Development, JenKem Technology USA, Eastman Kodak/Eastman Chemical Company (retired), Santa Rosa, California 95404-1585
[*]E-mail: mfrishberg@gmail.com

By identifying and responding to a critical, but unexpressed company need, a career envisioned to be spent entirely in the organic chemistry lab was put on a transition path into business development. This eventually led to exposure to the global pharmaceutical, agricultural, and industrial chemical marketplaces and much more chemistry and influence than staying in the lab – just not hands-on lab work anymore. This presentation will follow this unexpected career transition and highlight the skill set that qualifies most chemists to enter the area of business development, and what additional skills and outlook will need to be acquired along the way to be successful.

I am currently consulting in the area of custom organic synthesis and business development, and employed half-time as the Vice President of Business Development for JenKem Technology, USA, a very capable Chinese company specializing in PEGylation technology and supply of PEG derivatives from R&D through commercial scale. The discussion that follows, however, is based on my 28 year career with Eastman Kodak and Eastman Chemical Company, followed by eight years as President of Seres Laboratories, Inc., a FDA registered, R&D and cGMP kilo-scale, custom synthesis laboratory in Santa Rosa, CA.

I was pleased to learn that this ACS Presidential Symposium had been titled "Career Challenges and Opportunities....," instead of the usual "Career

Alternatives" or "Non-traditional Career Paths for Chemists," which is often used for this type of program. Traditionally, the mindset of many people has been that any chemical career outside of the lab or academe is an alternative or non-traditional career. I have always taken a more simplistic view that these distinctions are not necessary or useful. All of our education and experience ultimately sharpens our skills as problem solvers, no matter where that knowledge and experience might be applied. Undergraduates are learning how to learn about the fundamentals of their major field and testing themselves against problems that have already been solved. Graduate students and post docs are expanding those skills on problem(s) that have not yet been solved. After college, this knowledge base is applied to solving the problems that an employer needs to be solved, or for which a funding organization will support solving. And the next level is that of identifying those problems that need to be solved, or are worth solving, and providing the organization and direction for the solutions.

Career Beginnings

Anyone who knew me early in my career would be surprised that I ever left hands-on laboratory work. I have always liked solving challenging organic synthesis problems and designing new molecules and processes, and I still do. I have always enjoyed and found wonder in the basic day to day experiences along the way; the colors, the odors, the ability to grow beautiful crystals and to enjoy the sparkle and clarity of a newly distilled product. This excitement and appreciation for chemistry is why I pursued an education in organic chemistry in the first place.

My first twelve years in the lab after college provided me with a wide variety of experiences, including some basic business economic fundamentals. Initially, working for Eastman Kodak at its Tennessee Eastman Division Research Labs in Kingsport, TN, my lab work was directed to designing new and improved colorants for internal photographic and textile applications, and working in the lab on the gram to hundred gram scale. Most of this employed aromatic and heterocyclic synthetic organic chemistry, a perfect match for my college training. In this work, there were performance features that a new colorant molecule needed to meet, but the actual structure was not defined. Properties such as lightfastness, solubility, wavelength, bandwidth, and others were the targets and the goal was to design molecules to meet them.

After six years at this work, I was transferred into the Industrial Chemicals Research Lab, performing custom organic synthesis for external customers in the pharmaceutical, agrochemical, and industrial chemical industries, as seen in Figure 1. Here the focus was completely different. The exact molecule desired by the customer, and usually its analytical specifications, were strictly defined. It was our job to design a synthetic route that was amenable to scale-up to the multi-ton scale to be produced in the five hundred to three thousand gallon, or continuous process, plant equipment. The desired chemical needed to be produced competitively to meet the customer's price target and our internal profit expectations. Solving these synthetic problems usually meant going to the chemical literature, finding two to five existing routes, deciding that none of them would meet the scale-up and economic targets, and then designing a new route that would – usually under a fairly tight timeline. The initial lab work was at the gram to low kilogram level, but with the understanding that successful completion of the work and acceptance of our business proposal by our customer could lead to scale-up very quickly to the multi-ton or higher level. An example of the process scale change generally needed is shown in Figure 2. Also, whereas interactions during my colorant chemistry lab work were all internal to Eastman, or to Kodak in Rochester, NY, the custom synthesis work involved more internal interactions outside the research labs; with development, manufacturing, sales and marketing, and occasionally, with customers.

Figure 1. In the lab. (Courtesy of Eastman Chemical Company.)

Figure 2. Lab to plant scale-up. (Courtesy of the author.)

Setting the Stage for the Career Transition To Come

After a year or two into my custom synthesis experience, our lab and this business area hit a lull, where the value of the new projects coming in was not very high. We also had a low acceptance rate for our business proposals, even though they appeared to meet customer targets. As those of us in the lab did not interact with customers to bring in new business or in presenting proposals, we were unsure of the reasons, but felt a responsibility to help determine why and try to turn this around.

An opportunity came when I was assigned an interesting new project for one of the major pharmaceutical companies. Two weeks after beginning synthetic work on the target molecule we received notice that the customer wanted a different compound. The new target was similar in structure to the first, but would require a different starting material and process. Two weeks later the target molecule changed again, and again two weeks after that.

Wanting to do a good job, on time, but frustrated by not getting any explanation for the changes in target, or even if this would be the last change, I pushed for a face to face discussion with the customer. It took over a month to push against both internal and customer systems to get this arranged. There was major push-back from our sales people, who apparently had had some bad experiences with our technical people talking with customers, and the customer, at that time, did not allow its technical people to speak with vendors. Eventually, a visit was arranged. We were able to clarify the situation on my project and agree on the best target structure. Coincidently, my being present on the customer visit helped solve significant problems on two other projects; one where a member of my lab had been given the wrong structure as a target, and the other involving a complaint that our price for a second manufacturing campaign of another chemical building block was much too high.

Positive internal feedback within our sales and marketing organization on this trip resulted in one of the other sales people asking if I could put together a presentation of our company's chemical and equipment capabilities for an agrochemical customer whom he suspected had a project that could be a good fit for us. This led to another customer visit and a connection that immediately resulted in a multi-million dollar project for one of our large scale facilities.

After this trip, the word was out that it was safe to take me to visit customers, and someone out of Research could actually be helpful in the field. The next request was to pick an area of the company's chemical expertise and put together an expanded chemistry presentation. Several of us in the lab brainstormed and decided that our diketene chemistry and related unique equipment capabilities was a good choice.

Diketene chemistry on a large scale is only practiced by a few companies worldwide, yet its many derived chemical building blocks at the time were used in more than a hundred pharmaceutical products alone. From the basic methyl and ethyl acetoacetates, the range of this chemistry covered pyrazolone couplers for photographic applications and a variety of antibiotics, including the second and third generation cephalosporins popular at that time, the dihydropyridine calcium blockers for lowering blood pressure, and several of the anti-ulcer drugs. In the agrochemical area, this chemistry led to several of the pyrimidine based insecticides. And in the pigment area, these were key components of the yellow to orange acetoacetarylides and the red to magenta quinacridones used in a range of industries from printing to automotive coatings.

Over the next two years, I travelled with our sales people to make this presentation at over 80 customer R&D labs around the US. With this much travel, my time in the lab was limited, but I had the satisfaction of helping identify valuable projects for other members of our lab and helping establish and strengthen relationships with our customers. Without realizing or planning it, my primary function had shifted from lab work to business development.

The Final Step out of the Lab

During my customer visits while working out of the Research Labs in Tennessee, in addition to finding large scale projects that were the target of our business area, many smaller scale projects were identified. While our plant site did not have the smaller scale equipment suitable to take on these kilogram to multi-hundred kilogram projects, the business unit at Kodak in Rochester, NY that serviced the R&D community through their catalog of thousands of products did. I referred numerous projects their way, introducing them to new customers. Several years later when that Kodak unit reorganized and decided to put more emphasis on custom projects, my name surfaced as someone to consider to lead the effort and I was offered the opportunity to move to Rochester, NY; a move that took me the final step out of the lab.

A Temporary Tangent To Acknowledge the American Chemical Society

The American Chemical Society, and the professional development experiences that its volunteer opportunities can provide, should be acknowledged at this point for its part in preparing me to be more comfortable and effective in the customer and business interactions described above.

Just prior to these transitions in job functions, I had moved up the ladder of increasing responsibilities within the Northeast Tennessee Local Section of the ACS to that of Chair. Concurrently, I had connected with the ACS on the national level, being appointed to the ACS Younger Chemists Committee, and eventually it's Chair, after being a participant in several of the committee's chemical career counseling "Roadshow" programs on college campuses (Figure 3).

Figure 3. YCC Roadshow on campus. (Courtesy of the author.)

These ACS volunteer activities provided me broad presentation experiences; from speaking to college students and faculty, and to other ACS national committees up to and including the ACS Board of Directors. Assuming responsibility for the YCC Roadshow programs allowed me to expand my network, as I needed to contact other industry and government professionals to invite them to speak at these programs. This usually led to contact with a potential speaker's supervision to obtain their approval to participate. Experience was also gained developing and operating budgets and working on and leading teams before I had the opportunities to do so as part of my Eastman position responsibilities. In addition, the ACS provided a platform for pursuing my hobby of sharing and explaining chemistry to the public through demonstrations (Figure 4).

Figure 4. Chemical demonstrations for the public. (Courtesy of the author.)

Defining Business Development

Most industries have business development positions, in one form or another, blending a high level of technical and business competence, with communication skills, to accomplish growing their companies through some or all of the functions listed below. Some companies have also taken to renaming their sales force with business development titles, in an attempt to approach customer contacts that are not open to a sales person calling on them. If a business development job description mentions sales quotas or remuneration through commissions, it is likely a sales job.

Having come this far in laying the groundwork for my career transition, it is time to define business development, at least from the perspective of my chemical career. Simply stated, business development can be defined as a company function dedicated to growing a company's business in new ways, encompassing all or some of the following:

- Finding and evaluating new businesses and markets.
- Identifying needs for new products and/or services in new and existing markets.
- Growing existing products and services into new markets.
- Identifying and building relationships, alliances, and partnerships.
- Identifying new technologies and/or potential products for licensing or acquisition.

To achieve these objectives, business development usually involves these activities:

- Understanding your company's strategic intent.
- Learning your company's strengths and weaknesses.
- Market intelligence and customer interactions.
- Representing your company during customer visits and trade shows.
- Seeking and evaluating new project opportunities from both a chemistry and business perspective.
- Pulling together other company functions to get the job done as a team.
- Negotiating Confidentiality Agreements and contracts.
- Proposal preparation and presentation.
- Any other ways that can achieve growing the business.

Reviewing my career path by job title below, even though not expressly mentioned in the job description, those positions in bold type involved a significant component of business development in their responsibilities:

Eastman Kodak/Eastman Chemical Company
  Senior Research Chemist
    Dyes Research Lab
    Photographic Chemicals Research Lab
    Industrial Chemicals Research Lab
  Technical Assistant to Director, Market Management
  Regulatory Affairs Manager – FDA Products
  Director – Fine Chemicals Markets Development
  Business/Commercial Development Manager
  Business Development Executive
Seres Laboratories, Inc.
  Vice President, Business Development
  President and Member of Board of Directors
JenKem Technology USA, Inc.
  Vice President, Business Development

Going from a lab position to that of directing and then running a business, there was an increasing level of job functions and responsibility:

Responsibilities from the lab
- Perform chemistry and cost evaluations.
- Make recommendations and transfer technology.
- Assess company strengths and present them.
- Assess marketplace needs and move to fulfill them.
- Establish and build relationships.

Responsibilities in running the business
- Project selection and managing project evaluation.

- Setting prices and negotiations of Confidentiality Agreements, proposals, and contracts.
- Overall profit and loss responsibility.
- Strategic planning.
- Personnel decisions.
- Department or Company leadership.

Learning along the Way

In progressing through a career that takes one into a variety of job assignments, there are numerous opportunities for personal growth along the way. For a chemical professional, such as me, these involve growth in both technical and non-technical areas. The non-technical areas often are the more challenging ones. They do not always come naturally to the technical person who is generally an introvert by nature. An example is learning to be comfortable interacting with customers and other exhibitors at trade shows (Figure 5). Particularly where travel, customer and team interactions, communication across companies and cultures, and managing other professionals are involved; learning and performance in these non-technical areas often determines overall career success and satisfaction.

Figure 5. *Trade Show Exhibit 2007. (Courtesy of the author.)*

While I do not claim that any of the following are necessarily unique or profound, here are some key observations and thoughts accumulated along the way during my career.

Project Selection

Every project can be a good project, and every project can be a bad project – it all depends on how it plays out.

The key is to try to decide quickly and either forge ahead or shut it down. If moving ahead, make sure to have check-steps along the way.

This is a portfolio business where working with the right mix of customers and chemistry is the key. Sometimes you need to take on a marginal project to help out a customer in order to gain the visibility and reputation that will qualify your company for the better projects to come.

The most important thing to remember: projects come and go, but build your relationships to last.

Communication

Stay alert. Even with a great project, a great team, and a great customer, it is unrealistic to assume that significant communication problems will not occur during the course of a project.

Even on the most highly technical projects, communication is:

- The most important ingredient to insure project success.
- The most likely cause of project problems.

Listening is often the key communication skill to insure success. When communication breaks down and disagreements or finger-pointing begins, this is the time to stay calm and focused on the overall goal. By listening closely and asking leading questions, this is a time when you can learn what is really important to your customer and other members of the team, and be part of the solution and not part of the problem.

Communication across cultural boundaries takes the form of both words and actions. Americans tend to use a lot of idioms in their speech, which rarely translate accurately into other languages. Giving a "ball park estimate" or saying that you are "shooting from the hip" or "playing the devil's advocate" will usually result in blank stares or shock, rather than nods of recognition. Eliminating the use of idioms when I travel overseas or entertain non-English speaking customers and colleagues has been a continuing challenge.

When traveling overseas from the United States, learning at least a few words of the native language of your customer can really help. A simple "Bon jour" and "Merci" is appreciated in France, as is a "Guten Tag" in Germany, and a "Knee How" in China. In Japan, one still needs to embrace, not avoid, karaoke. Bonding with customers "on their turf", as in Figure 6, often means "being a good sport," two other idioms that do not translate well. Watch out for hand gestures, which often take on different and potentially insulting meanings in other cultures. Do not point your finger at anyone in Asia, or show the soles of your shoes by crossing your legs while sitting. Do not search out the US food chains and miss the opportunity to sample and discuss the local cuisines during informal sessions.

A lot of this is common sense, but the stress of travel and business pressures can take a toll on common sense.

Figure 6. You had to be there. (Courtesy of the author.)

Travelling

The ability to travel is an acquired skill!

Many technical people have steered away from potential career growth and successes due to a discomfort in travelling. Adjustments in sleeping, eating, and having a lack of control in ones surroundings can lead to a loss of focus and high anxiety for many people. On my first international trip, an engine on the plane caught on fire over the ocean on the way to France and we had to return to JFK airport. On the same trip, on the way back out of Heathrow in the UK, I was bumped from a confirmed business class seat with no explanation and had to fight to get re-booked and arrange for accommodations and meals due to the resulting overnight delay.

In addition to travel disruptions due to weather or overbooked flights, you will also have to be flexible and adapt smoothly to changes in plans brought about by business issues. Once when I was attending a chemical trade show in Amsterdam I received a message that I needed to adjust my plans and leave early to see if I could help solve a problem that a customer at a plant in France was having with one of our products. Fortunately, I was working for a very professional company who made the change in travel plans for me and provided one of our French field reps to help guide the way and act as an interpreter. One is not always so fortunate.

Business travel is rarely a vacation. It is part of your work, but it can be an adventure, rather than drudgery, if you approach it with the right frame of mind. Attitude is everything. And sometimes, when you are living in Rochester, NY and it is the middle of winter, as shown in Figure 7, a call from your West Coast Field Sales Rep requesting your assistance in visitng a customer is not an unwelcome call to get. Just check to make sure the kids have shoveled the driveway, Figure 8, before you return home.

The best travel advice that I can offer is never check your luggage; and bring along some work, a good book, a snack, and your sense of humor.

Figure 7. Winter in Rochester, NY. (Courtesy of the author.)

Figure 8. Winter in California. (Courtesy of the author.)

A good rule to follow is that if flight plans are falling apart while on a trip, and you have a chance to make a connection that gets you closer to your ultimate destination, take it. If you are running around an airport or train station with unsettled plans, don't pass up a restroom stop, as you do not know when the next opportunity might present itself. But if you grab the quick hot dog on the run, it will likely talk back to you the rest of the day. Meals on the road can vary greatly, from missed meals to those seen in Figure 9 and Figure 10.

Figure 9. Too much of a good thing. (Courtesy of the author.)

Figure 10. Where to start? (Courtesy of the author.)

The Pluses and Minuses of My Career Transition from the Lab to Business Development

For the laboratory oriented chemical professional, there is always a considerable amount of anxiety in any career transition to move outside of the lab. Throughout my career, my guiding principle, and one I share during career counseling sessions with students, is that sometimes one needs to get out of one's comfort zone in order to grow. From my experience, here are some of the pluses and minuses to consider if that move is into business development.

- +/- You can see and do more chemistry than if you stay in the lab, just not "hands-on" anymore.
- − Due to Confidentiality Agreements you cannot talk about most projects, even within your company, nor include them in external presentations or publications.
- + You get to know your company much better and have a chance to make a difference in many areas.
- +/- You get the added responsibility of being the key point person on projects, but need to be able to manage effectively with ambiguous authority.
- + There is the opportunity to grow one's network, often globally.
- +/- Travel can be exciting and broadening, but it can also be tedious, and cause you to miss family events.
- +/- There are chances to celebrate along the way, though often cut short by having to respond to the next challenge.
- +/- It is rarely boring, although the "to-do" list never ends.
- +/- There is the satisfaction of overcoming many problems, although there can be an endless supply of unexpected obstacles.
- + There is the opportunity to collect an amazing amount of stories.
- − Your best performance is often not visible to your supervision.

Looking back on my career, I would judge that the pluses have greatly outweighed the minuses, and dealing with many of the minuses has made me a stronger and more effective contributor. The greatest potential minus is the last one on the list. Depending on one's supervision and company culture, this is an area where one may need to pay attention and seek options, such as soliciting customer input, in order to receive the appropriate rewards and career advancement commensurate with one's accomplishments. One of the other minuses, that could be significant for someone who has the need to grow their reputation by publishing in scientific journals, is the restricted ability to publish due to having one's work covered under Confidentiality Agreements. Sometimes it is possible to get your name on a paper or patent, but it is more likely that presentations internally, and during customer visits, conferences, and trade shows will be your primary outlet, examples of which from my career are shown in References (*1–9*).

My time in business development positions has provided many career satisfactions. I have been involved in thousands of project evaluations and hundreds of active projects. But even more satisfying is knowing that my involvement has made positive contributions to the success of at least eighteen commercial drugs, medical devices, and similar products, including three blockbusters. Beyond that the value of these projects was worth over three hundred million dollars for my employers and billions of dollars for our customers, has been the knowledge that these contributions have made these products more economical and on the market months or years sooner than anticipated for people in need of these new medicines. Most of all, there has been the privilege to meet, work with, mentor, and learn from thousands of very special people worldwide along the way.

Conclusion - Job Description for Business Development

I do not know whether I would have responded to the following job description at the start of my career if it had been presented to me, but looking back this is what I bought into. If it appeals to you, give it a try.

Looking for a technically competent individual who is: personable, creative, entrepreneurial, highly motivated, results oriented, credible, decisive, resilient, flexible, ethical, responsible, a fast learner, trustworthy, a good listener, and a team leader and player, with excellent communication skills.

Must be able to travel or be audited at a moment's notice, deal happily with unreasonable people, and juggle multiple, ever-changing projects all at the same time.

Benefits include opportunities to earn frequent flyer miles, have good dinners with strangers, be the messenger who is applauded or shot, and get hands-on experience with crisis management, conflict resolution, and managing with ambiguous authority.

Control freaks and faint-hearted people who cannot tolerate rollercoasters need not apply.

References

1. Exhibitor Showcase Presentations, Informex Custom Chemical Trade Shows: Seres Laboratories, Inc., 2003-2008; Eastman Fine Chemicals, 1999−2001.
2. Tour and cGMP discussion leader, IQPC 6th Contract Manufacturing for Pharmaceuticals Workshop, San Francisco, CA, June 25, 2007.
3. Pacific Region Clinical Supplies Conference, What not to do when outsourcing APIs for Clinical Trials, San Diego, CA, April 22, 2005.
4. Successful Strategies in Drug Development Symposium, Scale-up, Manufacturing, and cGMP, San Mateo, CA, June 29, 2004.
5. Gaining professional employment, national and local ACS sponsored chemical career counseling presentatation for over 100 colleges and universities, 1980−present.
6. Confererence on Pharmaceutical Ingredients, Sharing the excitement - fine and specialty chemicals derived from 3,4-epoxy-1-butene technology, Turin, Italy, November, 1997, Hong Kong, June 1996, and at twenty pharmaceutical companies worldwide, 1996−2001.
7. Eastman Chemical Company internal sales force training, Your price is too *@!$$%??! high, Kingsport, TN, 1996−1998.
8. American Chemical Society National Meeting oral presentation, Strategic project selection in custom and fine chemicals, March 25, 1996.
9. Building blocks from diketene/ketene with pharmaceutical, agrochemical, and industrial chemical applications, presented at eighty customer R&D centers in the United States, 1981−1983.

Chapter 7

Connecting with a Most Powerful Ally of Science

George Rodriguez*

Argeni LLC, PO Box 63, Madison, New Jersey 07940, United States
*E-mail: cmewebcast@gmail.com

Throughout human history, chemistry has played a central role in developing new materials and innovative solutions to overcome societal challenges. At the same time, science, when working in concert with business, has been a truly powerful engine for prosperity. Proficiency in both disciplines is fundamental for professionals to be more productive, work more harmoniously and provide a better quality of life. This essay will discuss how growth in population and material consumption point to an expanding role for chemists and chemical engineers who also understand business. It will also highlight how the influence of government and media in society calls for the increasing participation of technologically savvy professionals in these sectors. Finally, examples will be given on how CM&E, the business and technology topical group of the American Chemical Society (ACS) New York Section, created new programs that have strengthened the support for science, technology, engineering and mathematics (STEM) education and helped ACS engage the public.

Introduction

In the coming decades the USA will face the insatiable global appetite for materials, the fall from the top spot in the global economy, and a political system unable to stop the relentless decline of its educational system or spur job creation. To change this gloomy scenario, the objectivity and problem-solving competence of STEM professionals need to be brought to the forefront of society. One tangible way of doing so is to increase the small percentage of lawmakers with STEM

© 2014 American Chemical Society

backgrounds to a larger number of authoritative voices to assure the USA's position as the largest innovation-driven economy in the world.

Also, there are glimpses of hope when States that focus on business growth succeed in education and STEM employment. To help the job creation process, the STEM community needs to break the academic silos so that students can expand their business knowledge. ACS has facilitated this effort with programs that include entrepreneurial mentoring. Its New York Section has successfully used its 60 years of connections with the business community to build an effective platform where students can learn about the rich mosaic of chemical enterprises and enhance their ability to shape the future of our world.

Population: The Last 500 Years Have Been a Blast!

In the nearly 200 thousand-year chronology of *Homo sapiens* (*1*), the last 0.25% of that period, or half a millennium, has left the most profound mark in the history of civilization. America was the last major continent on Earth to become part of the world map after the trip organized by Cristóbal Colón (a.k.a. Christopher Columbus) discovered land on October 12, 1492. At that time the global population was estimated at 500 million (compared to 300 million at the start of the Common Era 1500 years earlier). Since then the population has grown exponentially to 7.2 billion people (*2*, *3*) accompanied by a vertiginous advancement of knowledge.

The phenomenal human population increase to 14 times since the time of Colón would have been impossible without parallel developments in technology, trade and government. The fight for democracy around the world for the last two and a half centuries was triggered by the unfair exploitation of people, the infamous taxation without representation and the lack of competent leadership. During this period the USA became a beacon for freedom and economic opportunity. These principles helped catapult the USA into the world's largest economic and military power. The fight for independence was accompanied by mankind's remarkable ability to produce and distribute goods worldwide.

The shift to a thriving global economy was made possible by a growing middle class focused on providing future generations Science, Technology, Engineering and Mathematics (STEM) education. These disciplines have been essential to business and society in developing, manufacturing and delivering new compounds, materials, devices, products and techniques.

The Rise of Prolific and Voracious Populations

By the end of the twentieth century, the USA increased the consumption of non-fuel materials by a factor of seventeen from about 0.2 million metric tons (mmt) to 3.5 mmt according to the 1900-2010 US materials consumption study of the Center for Sustainable Systems of the University of Michigan (*4*). The USA population during that period increased by 233 million or a factor of four (from 76 to 309 million) and material consumption per capita also had a substantial impact.

The findings in the study concluded that "In 2000, U.S. per capita total material consumption (including fuels) was 23.6 metric tons, 51% higher than the European average. Construction materials, including stone, gravel, and sand comprise around three quarters of raw materials use. The use of renewable materials decreased dramatically over the last century—from 41% to 5% of total materials by weight—as the U.S. economy shifted from an agricultural to an industrial base. The ratio of global reserves over production rates is an indicator of the adequacy of mineral supplies; it can range from over a millennium (aluminum), to a few centuries (platinum, phosphate rock, chromium), to several decades (selenium, indium, cadmium). Several rare earth elements and other minerals critical to producing solar panels, wind turbines, and electric vehicles could face shortages during the next five to ten years."

The use of resources in the 20th century will pale in comparison to future needs. According to a UN report, by the year 2028 India will have 1.45 billion people, the same population as China, and the world population will reach a staggering 8.1 billion (5). China and India, combined, will represent 36% of the world's population and will have approximately eight times the population of the USA. As their middle class rises with the expectation for a better quality of life, their use of raw materials will intensify the scarcity of key resources.

Chemistry to the Rescue of a Material World

In the coming four decades, humanity faces an additional increase of nearly 3 billion people which will exacerbate the "Top 10 Problems for Humanity in the Next 50 Years" listed by Dr. Richard Smalley (6, 7), the 1996 recipient of the Nobel Prize in Chemistry, in May 2003: Energy, Water, Food, Environment, Poverty, Terrorism & war, Disease, Education, Democracy and Population.

The population growth and per-capita consumption will continue to have significant multiplier effects on the overall global material consumption. Economic advancement will require finding solutions to multidimensional problems. In the environment field alone society faces resource depletion, deforestation, desertification, animal mass extinctions and widespread pollution, as shown in the great garbage patches and ocean trash plaguing our sea (8).

Business and STEM education have proven Malthusian theory predictions wrong and they are poised to continue their success. However, in order to meet the multiple needs of humanity, all facets of society need to work in concert. Adding business skills to the profile of chemistry professionals will help society navigate the increasing complexity in managing the world's resource exploitation, transformation, delivery, recycling and replenishment.

Adapting to a New World Order

The geopolitical shift during this century will be of titanic proportions. The USA will have a smaller economy than other nations, China and India, for the first time in a century. Russia, with a territory that represents one sixth of the earth's land area, its vast nuclear arsenal and its sharp differences with USA's geopolitical interests, will continue to be an untamable force.

In a world that has been capable of cataclysmic mass destruction for 70 years, the potential increase in extreme political or religious beliefs accentuates the shadow of armed conflict and terrorism. The unpredictable behavior of nuclear North Korea and the recent military confrontations that involve three (China, Japan, and USA) of the world's top four economies due to territorial disputes (9) represent the tip of an iceberg that continues to test world stability.

In addition, technology is changing profoundly the nature of warfare and terrorism since the level of casualties once associated only with nuclear war can now be achieved by small-scale strategic attacks. The vulnerability of the USA electric power grid to an electromagnetic pulse (EMP), from sun flares or a nuclear explosion, or, physical (or cyber) attacks by terrorists like the one that took place by gunmen on April 16, 2013 in San Jose, California (10), has come to the fore. There are over 2000 high-voltage transformers in the USA grid with a single transformer supply lead time of one-year. The blackouts would be catastrophic and require a new set of cross-sector technology solutions.

Another factor is the rapid erosion of USA Competitiveness from the top position in the period of 2007-2009 to the 7th place in the 2012-2013 Global Competitiveness Report of the World Economic Forum (11), which adds to the imbalance of power. Finally, among all the multi-sector challenges, the most critical strategic threat is that the USA, an innovation-driven economy, has failed in a key factor for prosperity: education.

The STEM Decline Requires a Drastic Remake

The USA educational underperformance is best described in the Paris-based Organization for Economic Cooperation and Development (OECD) biennial study of global education systems (12). It showed that American 15 year-olds tested were mediocre, and rapidly declining, in reading and science skills as well as below-average in math. American students' rankings in math have slipped from 24th to 29th compared to the last test in 2010. In science, they've gone from 19th to 22nd and in reading from 10th to 20th. This is in sharp contrast to the surging performance from Asia (China, Japan, South Korea, Taiwan, and Singapore), Europe (Poland, Finland and Holland), as well as neighbor Canada and Australia and represents the persistent failure of not just our political and educational leaders but of our society and culture (13).

In the field of Chemistry, the results of the Annual International Chemistry Olympiad similarly show that the USA underperforms countries that spend a fraction of the USA investment in education.

Only 5% of USA college students graduate with STEM degrees which does not bode well in the future world stage considering that in China 46% of college graduates enter the workforce with degrees in these fields. STEM jobs in USA are expected to grow at twice the rate of jobs in any other field over the next several years (14) and many of these employment opportunities will not be seized by USA graduates.

By increasing business understanding and engagement, members of ACS, the largest scientific society in the world, will be more effective in getting corporate support of ACS programs including scholarships, student projects, mentoring, tools for teachers and other programs. Another key ingredient for STEM education success is the ACS member engagement with the media and the government, particularly in Congress and at the state education departments and board of education for deep changes may be required in school discipline, teacher training, STEM curriculum, laboratories and other areas.

Education, Business and STEM Jobs

The symbiosis between business and education is evident through the success of public schools in Texas and Florida which consistently ranked among the top 20 in the Washington Post 2013 ranking of the top 1900 public high schools nationwide (*15*), and the ranking of Texas and Florida as the top two States for business by Chief Executive magazine with a bimonthly circulation of 42,000 aimed at top executives (*16*).

In the 2018 STEM jobs projections by the US Bureau of Statistics, the State of Texas ranks second at 758,000 and Florida ranks fourth at 411,000 (*17*). Other states that rank well in future STEM jobs include California, Colorado, Maryland, Massachusetts, Minnesota, New Jersey and Virginia, among others.

Texas will have 61% more STEM jobs than the State of New York which ranked third at 477,000 but normalized for population size the difference is even greater in favor of Texas since its population is larger than New York's by only 34%. The State of New York is trying to catch up by attracting new businesses with substantial 10-year tax incentives. It remains to be seen if this will compensate for the lack of state income taxes in Texas (and Florida) and whether New Yorker taxpayers, who ultimately will be paying for these incentives, believe that it is fair to existing residents.

While studies are needed to determine causality and correlation over time, the preliminary findings suggest that business, education and STEM jobs can thrive together, at least in states with no income taxes.

Empowering the STEM Role in Society

The leaders of the world will face critical technological challenges in order to satisfy the largest population in history amidst a fragile geopolitical environment. This demands strong foundations of STEM knowledge to enhance the quality and timing of decisions. However, one of the most important decision making bodies, the U.S. Congress, which add up to 535 people, consists of only 5% of STEM educated professionals, including just two chemists. Nature Magazine quotes the STEM professional Bill Foster, PhD (Representative of Illinois) as a physicist who "wants more scientists in Congress who can bring to bear an analytical mind-set to lawmaking" (*18*).

The USA has accumulated a public debt of over $17 trillion ($17,000,000,000,000.00) a number of astronomic proportions. On October 18,

2013 the public debt surpassed that $17 trillion milestone with a $328 billion increase in one single day *(19)*. To keep that one day increase in perspective, we need to realize that it was larger than all the goods and services produced in one year (GDP) by any of 160 countries in the world which represents 83% of all 193 member states of the United Nations *(20, 21)*. Another point of reference is the total 2014 USA foreign aid budget amounts to $48 billion *(22)*.

The unfunded liabilities level in the US Debt Clock shows $127.5 trillion as of January 2014 *(23)*. The debt and liability figures are so phenomenal that few words can properly describe the geographical branch of *Homo sapiens* that has allowed this behavior to continue over the years. At this point, reversing these trends is not simply a matter of voicing a sudden sense of urgency. The understanding of basic STEM concepts and methods is fundamental in developing comprehensive plans and realistic timing, avoiding policy misdirection and minimizing waste of valuable resources.

Society must focus on increasing the level of overall STEM and business competence of the social fabric of the nation including academia, government and the media, also known as the powerful Fourth Estate *(24)*. ACS has been empowering members through its entrepreneurial support program, a laudable effort that shall contribute to the generation of STEM jobs.

Breaking the Academic Silos

A new breed of leaders with strong STEM foundations is needed in order to analyze, strategize and transform the multi-faceted world in which we live. Nanotechnology is a prime example of how academia can break the traditional boundaries in order to solve complex problems through the better integration of various scientific disciplines. Similarly, the inclusion of entrepreneurship in the academic training that complements STEM education embraces the concept of assimilation of new fields of knowledge that contribute to the mission of job creation while balancing resource management and environmental stewardship.

Six Decades of ACS Business Connection

ACS has one subsection with a six-decade tradition of business focus, the Chemical Marketing and Economics group of the ACS New York Section (CM&E). This group has become a focal point of influence for members of the technical, entrepreneurial and investment community in the business of chemistry. Founded in 1954 to address marketing and economic challenges in a comprehensive way, CM&E organizes monthly luncheons in the heart of New York City where thought leaders in energy, materials and life science share their original research and insights with a diverse audience that includes industry, investment, academia, media, entrepreneurs, general public and government at the state, national and supranational level *(25)*. Typical monthly attendance ranges from 30 to 60 people and over 150 people for the annual awards event.

Local and Global Outreach

Since 2009, CM&E has doubled luncheon attendance, increased its email list by a factor of seven and membership by a factor of ten. The group fund-raising in 2013 surpassed the contributions of all 58 prior years combined. Over half of CM&E membership comes from members of the public. The CM&E focus on innovation, operational excellence and having fun in the process, has resulted in initiatives designed to improve the support for STEM education, including the ACS Scholars program, and connections with the public and diverse non-profit organizations with global reach. The group has attracted the vast international community in New York and has nurtured a relationship with the Office of the Secretary General of the United Nations. Free webcasts are now offered to all ACS members. The Annual Leadership Awards established in 2012 celebrating the business of chemistry have been sold out. For the inaugural CM&E awards event, the New York Section received the ACS ChemLuminary Award for Global Engagement (*26, 27*).

Helping Students Become Future Leaders

CM&E has also instituted the position of Associate Directors of its Board (*28*) which allow chemistry and chemical engineering students in the Greater New York Area to participate in CM&E's organization of activities, gain knowledge of the broad economic impact of chemistry in the world, understand the latest market and regulatory trends as well as job opportunities worldwide.

In a survey of students who attended CM&E events, over 90% rated them at the highest level of satisfaction. "As a PhD student doing lab work every day, the business and industrial view of certain market is truly helpful for me to understand the full picture of one field," one respondent said when asked how the meetings will help his career. "As a student there is no academic outlet to learn about the structure of the chemical industry and get a grasp on how large and varied it is. CM&E topics definitely helped me understand this," wrote one student illustrating how monthly meetings provide a platform for mentoring students, increasing the connection with current leaders of society and sharpening their networking skills to enhance their placement after graduation.

Conclusion

As a preeminent innovation-driven economy, the USA relies on technological advancement and a high degree of business sophistication. In the coming decades, humanity will rely increasingly on multi-faceted and technologically competent leaders to navigate the world's ever more complex challenges that demand outstanding strategy inception and execution talent.

Business has been one of the most powerful allies of STEM education and together they have moved society forward to a level unimaginable just one century ago. Members of the American Chemical Society, such as chemists and chemical engineers, are uniquely qualified to follow career paths in business organizations and play a large role in society to bring the scientific methodologies to decision making across a variety of sectors.

The CM&E group activities of the New York Section represent a scalable template that ACS could use to increase its sphere of influence in local business communities in various centers of economic activity throughout the globe.

The American Chemical Society must continue its quest to engage the next generations of professionals with initiatives that combine STEM education, business and the leadership skills necessary for job creation and sustainable development. Fostering such programs throughout all major global economic clusters will help increase the transformational power of the society to advance education and quality of life, the essential pillars for peace and prosperity.

Acknowledgments

Thanks are due to Marinda Wu and Madeleine Jacobs for their leadership and inspiration to move STEM education forward and make it fun! I am also grateful to the various CM&E Board members, including Associate Directors, Anna Powers and Debra Rooker, for their support in reviewing the essay, to the leadership of the ACS New York Section Board for their guidance and to the Directors of CM&E who helped conceive and implement pioneering initiatives in a way that made them feel effortless.

Mention of government groups and other institutions in this publication is solely for the purpose of providing specific information and does not imply recommendation or endorsement by any organization.

References

1. *Homo sapiens, Human evolution evidence*. Smithsonian National Museum of Natural History. http://humanorigins.si.edu/evidence/human-fossils/species/homo-sapiens.
2. United Nations Population Division. http://www.un.org/esa/population/publications/sixbillion/sixbilpart1.pdf.
3. UN World Population Press Release of 13 June 2013. http://esa.un.org/wpp/Documentation/pdf/WPP2012_Press_Release.pdf.
4. 1900-2010 US materials consumption study of the Center for Sustainable Systems of the University of Michigan. http://css.snre.umich.edu/css_doc/CSS05-18.pdf.
5. *World Population Prospects. The 2012 Revision*; United Nations: New York, 2013. http://esa.un.org/wpp/Documentation/pdf/WPP2012_%20KEY%20FINDINGS.pdf.
6. Smalley, R. Presented at the Energy and Nano Technology Conference; Rice University, May 2003.
7. Funeral Service for Professor Richard Smalley Speakers: James Tour, Hugh Ross, and Ben Young, November 2, 2005.
8. Ocean Trash Plaguing our Sea, Smithsonian Ocean Portal. http://ocean.si.edu/ocean-news/ocean-trash-plaguing-our-sea.

9. *U.S. affirms support for Japan in islands dispute with China.* Reuters: November 23, 2013. http://www.reuters.com/article/2013/11/27/us-usa-china-idUSBRE9AQ0T920131127.
10. Smith, R. *U.S. Utilities Tighten Security After 2013 Attack*; WSJ. http://online.wsj.com/news/articles/SB10001424052702303874504579372990589502828.
11. 2012-2013 Global Competitiveness Report. World Economic Forum. http://www.weforum.org/issues/competitiveness.
12. OECD, Programme for International Student Assessment (PISA). Paris, 2013. http://www.oecd.org/pisa/keyfindings/pisa-2012-results-overview.pdf.
13. Wilson, B. Media and Children Aggression, Fear and Altruism. *The Future of Children*; Princeton-Brookings: Spring 2008. http://futureofchildren.org/publications/journals/article/index.xml?journalid=32&articleid=58&sectionid=269.
14. National Math & Science Initiative. http://www.nms.org/Portals/0/Docs/Why%20Stem%20Education%20Matters.pdf.
15. The Washington Post 2013 ranking of 1900 public high schools nationwide http://apps.washingtonpost.com/local/highschoolchallenge/schools/2013/list/national/.
16. Chief Executive magazine rank of 2013 Best and Worst States for business. http://chiefexecutive.net/best-worst-states-for-business-2013.
17. Texas' Federal R&D and STEM Jobs Report, 2013. http://www.stemconnector.org/sites/default/files/sbs/CVD2013TexasInnovation.pdf.
18. Nature News, November 15, 2012. http://www.nature.com/news/physicist-elected-to-congress-calls-for-more-scientists-statesmen-1.11839.
19. Washington Times article on the U.S. Public Debt, October 18, 2013. http://www.washingtontimes.com/news/2013/oct/18/us-debt-jumps-400-billion-tops-17-trillion-first-t/.
20. List of Countries by GDP. Wikipedia. http://en.wikipedia.org/wiki/List_of_countries_by_GDP_(nominal).
21. United Nations member states. http://www.un.org/depts/dhl/unms/whatisms.shtml.
22. The U.S. State Department Foreign Aid Fact Sheet. http://www.state.gov/r/pa/pl/2013/207212.htm.
23. Unfunded Liabilities in US Debt Clock. http://www.usdebtclock.org/.
24. Kronig, J. *The Tyranny of the Fourth Estate.* http://www.policy-network.net/uploadedFiles/Publications/Publications/Kronig_pn3.2%20p56-63.pdf.
25. Chemical Marketing & Economics group of the ACS New York Section (CM&E). About Us. http://cmeacs.org/aboutus.html.
26. 15th Annual ChemLuminary Awards, Global Engagement Award, Committee on International Activities. Indianapolis, IN, September 2013. http://vimeo.com/74288609#t=55m01s" target="_top".
27. CM&E Leadership Awards webpage. http://cmeacs.org/award2013.html.
28. CM&E Home webpage. http://cmeacs.org/index.html.

Chapter 8

Alternative Chemistry Careers: Using Patent Assets for High Technology Commercial Innovation

James L. Chao*

Retired Emerging Technology Strategist, IBM Corporation, and Adjunct Professor of Chemistry, Duke University, 7424 Ridgefield Drive, Cary, North Carolina 27519-0503
*E-mail: chao_j@bellsouth.net

The goal of this paper is to provide pertinent information on some of the career opportunities that are presented to a practicing chemist in the workplace and how it is possible to find an alternative career in business development and intellectual property law. For ease of presentation, this paper is divided into three sections. The first begins with a synopsis of my personal career as a chemist and how in industry one must also take an active role in preparing oneself for the benefit of the company. The second part introduces what is meant by IP, or *intellectual property*. I will discuss how an IBM team leveraged the IBM IP portfolio to create innovation and new businesses. The final part shows some examples of well recognized success stories about new business ventures based on that strategy.

Introduction

Early in my career as a chemist, I loved working in the laboratory. I thought that by joining a large high tech company like IBM that I would be able to work in the research laboratory until retirement. However, my managers insisted that career advancement would require that I grew out of the laboratory and instead work more closely with business executives. They needed thought leaders to

© 2014 American Chemical Society

help them figure out new ways to grow the business. After twenty years in the laboratory, I found that IBM was looking for technical people with wide-ranging expertise to become part of an Internet age think tank to explore opportunities for commercial innovation in businesses new to IBM.

In 2000 I joined a newly formed group called Strategic Intellectual Property Licensing. One of its primary goals was to use the roughly 50,000 active U.S. IBM patents and identify those patent clusters that could be packaged and licensed to other companies. This could result in forming a new collaborative venture or alternatively just cross-licensing IBM patents with theirs. In other cases, team members studied competitive products and services to determine if there might be possible infringement of the IBM patent portfolio. It became an important part of the intellectual property organization that contributed close to $2 billion per year to the bottom line for the sale of assets, or the licensing of patents or trade secrets.

As a chemist, I was charged with looking at patents outside IBM's normal business lines, such as in the industry sectors of biotechnology, pharmaceutical, petrochemical, polymeric materials, and medical devices. Since IBM has important customers in each of these industrial sectors, the strategic licensing team was able to talk under friendly terms to many companies about exploiting these patents and other intellectual property assets. In this chapter, I will also discuss a few of these innovative business ventures that became commercial success stories that resulted from the IBM patent portfolio.

Part 1. Personal Background as a Chemist

The career of the practicing chemist involves a series of decision points as chemistry is a very broad field and there are many ways that a professional has to make some important choices. My career decision points are shown in Figure 1. My personal goals meshed with the goals of my company and prepared me for the opportunity to work in the business development and intellectual property area

1. My parents were trained at Purdue University after World War II, my father as a chemical engineer and my mother as a biochemist. Like me, both of my siblings obtained chemistry degrees from the University of Illinois in Champaign-Urbana. One brother is a retired chemical engineer and the other is a family practice physician. I always wanted to become a chemist.
2. Because of this determination to be a chemist, I wasted no time in getting through my academic training. I got my B.S. in 3 years, and then studied an additional year to obtain a Master's degree. I was the first to finish in my class at UC-Berkeley, completing my Ph.D. in 4 years.
3. As an undergraduate, I took all the chemistry classes available and early on thought how cool it would be to be a theoretical organic chemist who knew lots of different ways of synthesizing new molecules.

However, when I was invited as an NSF intern to work in the laboratory of a physical chemist at Yale University over the summer of 1974, I immediately decided that experimental physical chemistry was what I wanted to do.
4. But I also had many outside related interests, particularly watching my roommate who was a Physics major. I volunteered with computer science professors to work on various projects and decided to teach myself digital electronics which led me to build microprocessor and computer controllers for chemistry professor's needs.
5. I decided to go to graduate school in physical chemistry worked for Professor Charles B. Harris on low-temperature solid state energy transfer problems. I built a number of microprocessor-based controllers for the optical multichannel analyzer (OMA) and was able to consult with a number of national laboratories to build systems for them as well while in graduate school.
6. Upon getting my Ph.D., I was torn between working in academia or for a top-level industrial research laboratory. I had offers from the University of Rochester, to work at the Laboratory Energetics, as well as Bell Laboratories, but ultimately decided to join IBM after speaking at their Thomas J. Watson Research facility in New York.
7. Several IBM managers informed me of a newly formed secret venture and they offered me a position where I would be able to be an FTIR applications scientist and be able to work freely with customers on ground breaking research throughout the U.S. This venture was later announced as IBM Instruments, Inc., a wholly owned subsidiary of IBM that sold research instrumentation using IBM developed computers, patterned after the Hewlett-Packard and Perkin-Elmer's business model.
8. I had a number of successes at IBM Instruments, Inc. including introducing the first FTIR run by a personal computer which was showcased at an IBM stockholders meeting in 1984. However, by 1987, IBM was no longer in a position to support this minor venture. A decision was made to dissolve IBM Instruments and sell it to a number of leading instrumentation companies. By then, the IBM Personal Computer had become ubiquitous in the instrumentation field and there was no longer a need to use instrumentation as a pretext to sell computers to scientists and engineers.
At about the same time, there was a movement in Research Triangle Park to create collaborative relationships with the research companies because of the close proximity to University of North Carolina, North Carolina State University, and Duke University. Duke which by then had appointed me with a visiting adjunct professorship appealed to IBM to transfer me to Research Triangle Park (RTP) and I became one of the first examples of the Technical Interchange Program. With the support of Dr. John Cocke, who was an IBM Research Fellow and a member of the Duke Engineering Board of Directors, the offer was made.
9. IBM prepared a contract for me with Duke University that allowed me to spend 40% of my time at Duke University and to share equipment from

my laboratory in RTP for use by students on their research problems. I became an adjunct associate professor and assisted in teaching Analytical Chemistry classes at Duke. I performed as a collaborator with a handful of Ph.D. candidates with Dr. Richard A. Palmer.

10. As part of the relocation, I was offered a position in IBM's Materials Engineering Laboratory in RTP, with primary responsibility for Environmental Corrosion Testing of IBM's hardware products. This work involved accelerated corrosion testing in pollutant gas environments for indoor and outdoor exposures.

11. I gained a reputation for publishing many internal IBM technical reports and external publications on my novel FTIR methods performed with graduate students at Duke University. In both 1990 and 1991, I was awarded IBM-Research Triangle Park's Outstanding Co-Author of the Year recognition. Shortly after that time, the Award was no longer made, and efforts were made to instead encourage researchers to generate patent filings instead. This was a major change for me and followed the direction of most industrial firms to not share technical findings and instead to protect their intellectual property

12. With the effort to file patents, IBM quickly became the top company to be awarded patents in the U.S. Since that time, IBM has consecutively been the leader in patents, with over 50,000 active U.S. patents. The timing in 2000 was perfect, since I had been looking for a change in career that would put me in close proximity to various business-line executives, and the company was assembling a think tank for leveraging the wide patent portfolio it possessed. The formation of the Strategic Intellectual Property group of technical experts offered me a chance to work outside of the chemistry laboratory for the first time.

The twelve listed career decision points led to a natural evolution towards working with inventors in the company to develop business ventures surrounding intellectual property and to generate large sums of patent licensing fees. In addition, I had a "can-do" attitude from my earliest days, often volunteering to do things outside of the normal realm of chemistry. I was able to leverage my outside interests in computers, first to design and build instrumentation in graduate school for chemistry professors, and later to join an instrumentation company, writing software and assembling electronics, and creating novel optical accessories. With this attitude, it was no surprise that I also became an adjunct professor at Duke, and was heavily involved in volunteering for the American Chemical Society and being elected an ACS Fellow in 2013.

In the next section, my transition to intellectual property law and my duties using my chemical and scientific expertise will be covered.

Dr. Chao's Career Decision Points

Undergraduate Career

1. Becoming a Chemist–entire family worked in chemistry
2. Determined to complete degree requirements for Ph.D. as soon as possible
3. Should I be more theoretical or experimental?
4. Was enamored by physics and computer electronics. Taught myself digital electronics and use of computer microprocessors for controlling electronic devices
5. Decided to go to graduate school to become a physical chemist

Graduate Career

6. Upon graduating from UC-Berkeley I had to choose between being a professor at University of Rochester's Laboratory for Laser Energetics, Bell Laboratories, and IBM Research
7. Ultimately decided to join a secret venture called IBM Instruments, Inc. as an FTIR Applications Scientist in 1980

IBM Career in Industry

8. After the dissolution of IBM Instruments, Inc. in 1986, I was invited to participate in the IBM's Technical Interchange Program (TIP) program with Duke University in Research Triangle Park, NC
9. The TIP at Duke afforded me 40% of my time as a Research Adjunct Professor
10. Also joined IBM's Material Engineering laboratory as a materials chemist, with primary responsibility for Environmental Corrosion testing of IBM's hardware components
11. To Publish or Not To Publish
12. IBM becomes the #1 assignee for the U.S. Patent and Trademark Office, a distinction kept for over a decade.

By 2000, I had several patents pending and my manager recommended that I look for a position outside the laboratory where I would gain more exposure to the executive ranks.

Figure 1. Career decision points for the author

Part 2. Career Change: Leaving the Laboratory

As stated earlier, I was informed that career advancement in industry was much different from an academic career, and that visibility to the business-line executives was paramount for promotion. While most employees advance to the top ranks by climbing the management ladder, a few achieve success on the technical ladder, but such success typically requires contributions that help the company generate very large sums of money. When I learned that they were looking to assemble a think tank of technical experts to work specifically on creating new businesses or licensing of the company's intellectual property assets, I responded to the calling. The new role required learning IP law, business development practices, and developing negotiating skills to convince executives and attorneys of potential joint ventures using intellectual property assets. To ensure that this new patent licensing group came up to speed quickly, we received training in intellectual property law and were given access to the IBM patent portfolio and licensing databases. We teamed up with intellectual property attorneys to scrub patents owned by IBM and those patents which potentially read on IBM products from other companies. We practiced arguing specific patent claims before attorneys and were teamed up with business negotiators within the IP organization. I also became involved in creating 90-day trial software licenses which allowed us to protect our intellectual property interests for early releases of future products. Later on, IBM paid for additional training from the Practising Law Institute (PLI) for a course given at Suffolk University Law School in Boston, as preparation for the patent bar exam.

This new intellectual property group was added to the existing IP organization, which was tasked to generate nearly $2 billion in special income each year.

Intellectual Property Licensing as practiced by IBM has proven to be a very lucrative source of income which requires an organization that efficiently develops these high-value assets which are comprised mostly of patents and trade secret know-how. Two approaches are organized in the IP licensing organization, one which is predominantly strategic and the other tactical. See Figure 2.

Intellectual Property Licensing

- Strategic vs. Tactical ➔ Good Cop vs. Bad Cop Approach
- Requires Knowledge of the State-of-the-Art Technology
- Requires Approaching Other Companies That Have Significant and Proven Financial Strength

---

Figure 2. Intellectual Property Licensing Approaches at IBM

The next two figures summarize the differences between the Bad Cop approach (Tactical) and the Good Cop approach (Strategic). In the Bad Cop approach (Figure 3), products and documentation from companies are analyzed against subject matter patents looking specifically for technology which may be infringing IBM's patent portfolio. If infringement is likely, then proof packages are created for consideration by the company's intellectual property attorneys. Further, a business assessment is made as to the business relationship with the company and the likelihood the ability of the offending company to pay damages or to exchange cross licensing fees if their patents may be ones that IBM could use effectively in its own product development efforts.

Bad Cop Approach

- Involves Using Proof Packages to Notify Companies of Possible Patent Infringement
- Use Patent Claim Language to Show That in Company Product Literature and/or Product Design That They Are Employing Patent Art Without License
- "Fair And Equitable Licensing" - Percentage of Revenues of Products Based on Number of Patents or Collection of Patents
    - Other Options: Cross-licensing Agreements Between Both Companies

---

Figure 3. Bad Cop Approach to Intellectual Property Licensing at IBM

In Figure 4, the Good Cop approach is shown. This approach requires examination of companies that are not in the computer/electronics industry sector, but that may have the ability to absorb IBM intellectual property assets in innovative new product areas. Depending on this business and technical assessment, a determination as to whether involvement in a joint venture including IBM experts would be necessary. Because so many companies are already customers of IBM, the customer relationship executive helps to grant us an audience to pitch the potential venture to the CEO, CFO, VP, or most often, their Chief Technology Officer (CTO) at the prospective company. A storyboard presentation is created, and the technical member rehearses the pitch and prepares to answer likely questions from the executives.

> **Good Cop Approach**
>
> - Creating Storyboards Based on a Joint New Business Venture Surrounding IBM IP
> - Need to Make Sure Ventures Are Fiscally Possible and the Other Company Has the Technical Breadth to Absorb and Use the IP
> - Are They the Best Company? Requires Investigating Their Patent Strengths and Willingness to Talk Under Confidentiality Agreements
> - Other Options: Royalty Payments

Figure 4. Good Cop Approach to Intellectual Property Licensing at IBM

I was most actively involved in the Good Cop approach because of my technical expertise as a chemist mostly closely aligned with companies that were not direct competitors to IBM. Figure 5 summarizes the division of industry sectors in IBM's case. The industry sectors which matched my expertise were in the petrochemical, pharmaceutical, medical devices, biotechnology, and applied materials sectors.

> **Industry Sectors**
>
> • **IBM Participating Sectors** – Computers, Semiconductors, Networking, Software, Digital Business Methods
>
> • **Available to a Chemist** – Petrochemicals, Pharmaceutical, Medical Devices, Biotechnology, Applied Materials
>
> Virtually All Companies Have Relationship With IBM

Figure 5. Division of Competitive Industry Sectors for IBM

Before a proposal can go out the door and a meeting is arranged with a prospective company, it requires considerable internal vetting. This process is summarized in Figure 6.

> **Meeting with Prospects**
>
> A Proposal is vetted internally with approvals from
>
> 1. Legal
> 2. Business Development - Negotiators
> 3. Technical Experts
>
> allowing the IP group to meet with Chief Technology Officers (CTO) or CEOs, CSOs, VPs

Figure 6. Requirements for Meeting with Prospects for a New Joint Venture

Part 3. Examples of Prospective Out-of-Industry Commercialization Ventures Using Intellectual Property Assets

In this final section, several examples are presented where intellectual property assets, including patents, were used to engage prospective companies in commercialization. All of these examples require developing a risk/benefit analysis as they almost always involve spending millions of dollars to fully develop. These examples show the value that the chemist can find an alternative career by thinking outside of the box and using the company's intellectual property assets to create new value to the company. I have purposely omitted the names of the other companies, the size of the deal, and other proprietary information.

These examples are taken from my knowledge of IP deals with other non-computer companies in my position as a strategic IP licensing program manager. As the subcommittee chairman on non-ACS National Awards in the ACS Committee on Patents and Related Matters, I was charged with finding chemists worthy of being recognized for the National Inventors Hall of Fame, the National Medal of Technology and Innovation, the National Medal of Science, and the National Women's Hall of Fame. I received support from the IBM Research Business Development office and coordinated my ACS effort with that of IBM's, getting letters of support from Dr. Paul Horn, Head of IBM Research at the time.

In the first example, IBM has proven expertise in the semiconductor manufacturing area making complex computer wafers with ever improving yields. Much of this success is attributed to the development of tools which are deployed during manufacturing in order to increase the yield. One of these tools employs complex mathematical computations and multivariate analysis of various input sensors in the steps of wafer manufacturing. A synopsis of the proposal to use this tool in the oil and gas exploration industry is shown in Figure 7. In this case, to demonstrate feasibility, post-analysis of drilling data were requested so that the multi-variate correlation could be done, prior to implementing the tool in real-time at considerable higher cost.

Petrochemical Example

- IBM Developed a Tool Used in Semiconductor Manufacturing to Improve Yield for Working Wafers
- Software Tool Employs Multivariate Correlation on Process Inputs During Recipe Steps to Improve Working Wafers
- In Oil and Gas Exploration, This Tool Could Be Employed to Improve Output Yield

Figure 7. Petrochemical Industry Example for Licensing of Software Tool from Semiconductor Manufacturing Processing

In Figure 8, another example of a proposal was made to amortize the costs involved in building semiconductor foundries in the billions of dollars, by figuring out a use for these when the semiconductor industry undergoes its unforeseeable over- and under-capacity cycles. This idea was pitched to players in the biotechnology and/or pharmaceutical industry that also have high costs associated with searching for new drug prospects. The proposal involved developing new methods for manufacturing gene chips or biochips.

Pharmaceutical/Biotechnology Example

- Semiconductor Industry Involves Extremely Expensive Foundries Which Experience Cyclical Demand
- In the Low Demand Cycle, Could These Foundries Be Redeployed to Make "Gene Chips" That Could Be Used to Find New Drug Prospects?
- Capabilities in Micro Fluidics, Pattern Imaging, and Packaging Would Be Valuable to This Nascent Business

Figure 8. Pharmaceutical/Biotechnology Example for Manufacturing Gene Chips in a Semiconductor Foundry

In Figure 9, several examples of patent assets and trade secret know-how which were pitched in the applied materials areas. The first was from patents I developed myself, which addressed problems with spilling of fluids on keyboards and keypads. The superabsorbent materials used in children's diapers could be used in keyboards to offer spill-proof protection.

Applied Materials Examples

- Superabsorbent Gel Materials Used in Diapers Could Be Used for Making Spill-proof Keyboards
- Self-assembled Monolayers - (SAM) Could Be Used For High Performance Marking of Progressive Lenses
- Electro-chemical Thin Film Hard Drives and Lubes Improve R/W Information Density and Reliability

---

Figure 9. Applied Materials Examples for Commercial Improvements to Manufacturing in non-Semiconductor Industries

In another example, manufacturers of optical eyewear use laser marking methods to manufacture progressive lenses. By instead using an inexpensive technique using self-assembled monolayers (SAM), costs and reliability could be improved.

Finally, electrochemists from IBM Research created methods for using thin-film heads which greatly increased R/W sensitivity and is now used by all hard drive manufacturers for its lower cost and much higher information density. Until this development, hard drives were huge by comparison. Furthermore, the use of di-electric materials and improved lubes have allowed for the enormous improvements in storage capacity in the industry.

In Figure 10, intellectual property assets were used for commercialization in the Medical Devices area of the Healthcare Industry.

In the first example, it is known that the U.S. population will suffer a lot from the hardening of arteries that leads to heart attacks and the post-operative need for stents for an aging population. Stents are primarily made from metal wires and are placed in damaged arteries to keep them open. IBM was approached because of its ability to manufacturer very fine metal structures in its semiconductor industry in a way that they could be mass produced in a variety of shapes and sizes. Using expertise and trade secrets from semiconductor manufacturing, a series of patent filings were made for making these stents reliably and economically. Since this technology differs greatly from the existing manufacturing methods, these patents were offered to various medical device makers already in the industry. It was proposed that these micro fabrication methods could also offer protection from restenosis and could incorporate drug eluting capabilities sought by the industry.

Medical Device Examples

- Patents Issued for Manufacturing Wire Stents Using Semiconductor Methods Used for Cardiac Patients
- Royalty-based License for Use of Excimer Laser Used In Lasik Eye Surgery
- Medical Robotic Surgery Patents Licensed to IBM for daVinci Robot Incorporating a Pre-determined Surgical Plan and "Brakes"

Figure 10. Medical Device Examples in the Healthcare Industry Resulting in Innovation Using Intellectual Property Assets

In another example, an IBM researcher was responsible for inventing a use for a new excimer laser to be used as the basis for LASIK surgery. This concept has been a commercial success resulting in royalty payments whenever the procedure is used for LASIK treatment to obviate the need for glasses.

Finally, IBM was requested by the Navy to develop means for conducting robotic surgery on board Navy vessels without requiring the surgeon to be there. Software methods were developed which allowed a robotic arm equipped with endoscopes to perform surgery remotely. The software allowed the surgeon to create a "surgical plan" which was automatically followed and offered many safety features, such as "brakes" to avoid mistakes during the procedures. These seminal patents are licensed to manufacturers today. They use these in robots primarily to perform hospital surgery with much greater precision on subjects where the procedure is routine and avoids many unpredictable complications.

In all these examples, intellectual property assets were used as the basis for innovating and creating novel solutions in commercial applications. They all required additional development to realize their final product successes, so it was not as if final solutions were being sold. Oftentimes, university researchers fail to realize that patents in it of themselves are not valuable unless someone is willing to spend millions of dollars to create the product. However, the exclusivity given to patent holders is oftentimes worth the risk and results in a lot of success.

Conclusion: Alternative Careers for Chemists Using Intellectual Property Assets for Commercial Innovation

The above examples show that the chemist can prove to be far more valuable to his company for focusing on intellectual property issues. All that is required is sufficient chemistry and scientific expertise and a willingness to think outside-the-box. Figure 11 summarizes these concepts.

> ## Conclusion
>
> - Chemists can be more valuable to their companies by developing business interests that take advantage of their technical expertise.
>
> - **Patents** are not the goal, but instead should be viewed as "starting points" for value to the company.
>
> - **Innovation** that creates new products and processes is the true objective and patents serve as one means to ensure exclusivity.

Figure 11. Summary of some of the concepts given in this paper

It is hoped that this paper shows that a rewarding career is possible outside the laboratory for the bench chemist, and that knowledge of chemistry is useful background for generating new product innovations when working in the intellectual property field.

Chapter 9

And then the Wind Changed

John Fraser*

Office of IP Development and Commercialization, Florida State University, 2020 Westcott Building North, 222 South Copeland Street, Tallahassee, Florida 32306-2743
*E-mail: jfraser@fsu.edu

I was newly married, living in Amherst, Massachusetts, after laying out for my wife a rosy future with me as a faculty member and then maybe university President. (And the wind changed.) We're in Berkeley; I'm getting a Ph.D. in biochemistry. (And the wind changed.) We're in Ottawa, Canada; in the Canadian counterpart to the NSF as Director of Programs. (And the wind changed.) We're in Vancouver, Canada; I work as a venture capitalist. (And the wind changed.) Now we are in Tallahassee Florida at Florida State University, where I was elected President of our professional association, AUTM and travelling the globe "communicating the value" of university research to make a Better World.

As a result, I have had a very eclectic a career - private sector to government to universities and back. Never underestimate the value of a good degree, but you need to always be on the lookout for interesting opportunities. Such an opportunistic approach does not make you fabulously wealthy, but it's worthwhile.

It is a pleasure to be writing about about career advancement in a rapidly changing world. This chapter will be less about strategies for career advancement and much more about lessons I learned personally that I hope will be of use and a reassurance to you in the years ahead.

The upshot is that by taking full advantage of every opportunity presented to me, both the large and small, the successful and the not-so, I have had a very

© 2014 American Chemical Society

eclectic career moving from the private sector to government to universities and back. Never underestimate the value of a good degree, but you need to always have your personal radar set for interesting opportunities. Such an opportunistic approach may not make you wealthy, but it's never dull, you'll have stories to tell and you can make a difference.

Amherst, MA

- Graduate student
- Movies: Butch Cassidy & the Sundance Kid
  Midnight Cowboy

- Lesson: Reach for the best opportunity

Keep in mind that it is very possible that you will be doing something in 10 years that you never thought you'd consider *or that doesn't even exist yet*. Keep your eyes and ears open always and enjoy the ride!

This is not about being blown across the country by chance, but about Seizing Opportunities

I've listed popular movies to give you a timeframe. I was attending UMass in Amherst as a grad student, newly married and living on almost nothing (you all remember how that goes!). Of course, I laid out for my new wife a rosy future for us in a linear manner after my Chemistry degree, as a bench scientist, as a faculty member and then maybe university President and living in an ivy covered house provided for by the university. *And the wind changed and we moved to UC Berkeley to accept the grad school offer I had initially turned down.*

LESSON: *Reach for the best opportunity as you have nothing to lose except a future wasted wondering "what if".*

Berkeley, CA

- Graduate student
- Movies: Diamonds are Forever, Dirty Harry

- LESSON: Make a decision, don't drift

108

We're in Berkeley, California; living in married student housing with a new baby. Going out consists of McDonalds and a drive to the beach. I'm working on a Ph.D. in biochemistry and doing rather well, I might add. *And the wind changed and I leave the PhD program with a Masters degree.*

LESSON: *I wanted to be to see if I was as good as the best. Yes, I was. The very best places force you to confront things and make decisions. I decided that I did NOT want to be a scientist building a perfect brick of knowledge to contribute to the world's general store of knowledge. I decided that I wanted to work with people who are using their science to take those bricks to build a Better World (whoops, a plug for AUTM's annual publication). So we left academe and the PhD program and moved back to Ottawa, Canada, with the lofty ambition of giving something back to my native country.*

Ottawa, Canada

- First Job – NSF counterpart
- Movies: The Exorcist, American Graffiti

- LESSON: Do things that distinguish you from the next person
- LESSON: Understand the viewpoint of others at all times

LESSON: *While in Ottawa, one day, my boss asked me: Do you know why I hired you? My answer: Good degree? No, because in your resume you explained how you organized a beauty contest while an undergrad as a fundraiser and I realized that you had the ability to deal with skittish, egos running around trying to best each other. Just what I need dealing with the academic community. Do things that distinguish you from the next person!*

LESSON: *As Agency Program Manager I signed 400 Letters with Reviewers Comments to grantees per year. I signed a lot of Letters but for each recipient there was only one Letter received, talking about the quality of their research. For Me – High Volume, Low Impact. For the PI, Low Volume, but very High Impact.*
LESSON *– try to understand the viewpoint of others at all times.*

I loved this government job. I was respected as the Messenger with Money. Eventually, I wanted more hands-on involvement in projects. So I saw an advert where I could use my education and network in a venture capital fund being started. I called people I knew in the city of the Job and found that they were involved in the Fund.

LESSON: *It really is a small world. Use your networks. Always be up for a coffee, lunch, dinner with colleagues or new acquaintances at conferences or seminars. It will surprise you how people will offer help, if you ask. Remember to thank people - I took time to figure out how to send my 2 references bottles of Champagne before FEDEX existed in Canada. A simple hand written thank you note stamped and mailed pays dividends in this computer age.*

> ### Vancouver, Canada
>
> - Venture Capital professional
> - Movies: The Empire Strikes Back
>   The Blues Brothers
>
>
>
> - LESSON: Sometimes S--t happens. Not your fault. Move on.

Next we're in Vancouver, Canada – lovely neighborhood right near UBC; I work as a venture capital professional. Kids are all growing like weeds which may have been due to the 300 days straight of rain, but…*And the wind changed and the company folded.*

LESSON: We started and grew a company, but recession hit in Canada and the two corporate shareholders were unhappy with the lack of financial return due to the often not recognized long product development timeframes involved. We were all let go, nice severance and the shareholders moved on. LESSON: Sometimes you try with everything in you, do a great job and circumstances conspire against you. Not your fault – pick yourself up and move on.

> ### Toronto, Canada
>
> - Consultant
> - Movies: Return of the Jedi, The Big Chill
>
>
>
> - LESSON: Network, Network, Network

And the wind changed and I became a consultant in Toronto. Several times in my career, I have been a consultant between formal employments. It pays the bills and that keeps my wife happy.

LESSON: As a consultant you will surprise yourself by how much you actually know and that there are people willing to pay for that knowledge. Always accept a job if you think you can do 50% of it; you'll learn to do the rest. Your skills will never match the requirements exactly, but if your interest is piqued and you "think you can", go for it. As my wife says, "They hardly ever throw real rock or tomatoes"; so, what do you have to lose?

LESSON: Always deliver a bit more than what you promised or earlier than promised and be willing to consider doing a small demo / 'freebie' to earn the business. Often that gesture is the leg up on securing that job. I've learned that there is rarely a front runner who is light years ahead of the pack – hiring often comes down to a coin toss and anything you can do to put your strengths in front of the decision makers is fair game.

LESSON: Being a sole consultant working at home is very lonely. And as my Wife told me one day: I married you for better or worse, but not for lunch. This is another example of how your contacts and networks will keep you sane.

We were ensconced in Toronto, owners of a house and three kids (one in a private school) and the Wind Changed.

Chapel Hill NC

- Company co-Founder
- Movies: Top Gun, 'Crocodile' Dundee

- LESSON: You can sell anything once, the real art is turning that sale into a sustainable business

We moved to the Research Triangle Park in North Carolina to start a venture capital backed company to provide tech transfer services to universities. This was six years after the technology transfer law known as the Bayh Dole Act was signed.

LESSON: We were running a service business transferring university technology to companies to develop products. I learned that in the US, you can sell anything once and someone will buy it. The real art is in growing that first sale into a sustainable business. I'm glad to say that we were successful and the business grew. After 4 years we sold it and money was made by all (just not enough to retire on). I restarted consulting and ... the Wind Changed ... we moved to Calgary, Canada, to replicate the service business to serve the University of Calgary.

Calgary Canada

- Company co-Founder
- Movies: Pretty Woman, Dances with Wolves

- LESSON: Kenny Rogers: "Know when to hold 'em; when to fold 'em and when to walk away"

We are in Calgary, bought an acreage on the side of a mountain, new mortgage, one kid in college, two in junior high, foster parents to 4 additional kids. The company where I was President, was a public private partnership and there were constant discussions as to whether this was a service company, or one that paid profits to shareholders. Team building was critical and teamwork paramount. After four years and having successfully developed a smoothly running team, I threw myself on the sword and voluntarily left mostly due to fatigue (the back-biting here was off the charts) and a Canadian recession that led to seemingly endless Board "do good or make money" arguments.

LESSON: Kenny Rogers said it best, "know when to hold 'em, know when to fold 'em and know when to walk away."

Blind River, Canada
- Economic Development Bank President
- Movies: Pulp Fiction, Forrest Gump

- LESSON: Things were not what they seemed. Cut Losses and Move on

And the Wind changed. We moved to Blind River Ontario to head up a well-financed Economic development bank in a distressed uranium mining community. The business idea was to financially back small businesses built up by the local Indian Band, and the small business community.

Raise your hand if you know where Blind River Ontario is. North Shore of Lake Huron.

And we hit a major bump in the road in a very small town. On our son's first day in school, I escorted him to the principal's office to get signed up. I was in a suit and tie and trench coat. The student body assumed I was a narc and our son was a plant, and you can guess where that went. My wife says no one spoke to her the entire two years we were there unless they were giving her change. The scenery however was magnificent and if you like snowmobiling, it's heaven!

LESSON: Small town living is not big city Living. We tried it for two years and finally gave up. We had not done our homework on the community, nor the Job. The Job sounded good on paper, but the reality was light years removed. So you cut your losses and move on.

> **Tallahassee, FL**
> - Founder of FSU tech transfer/commercialization office
> - 2006 - President of AUTM professional Society
> - Movies: Pirates of the Caribbean II, daVinci Code
>
>
>
> - LESSON: The World is not a foreign place. People raising families in Vietnam and India desire the same things we do for the future – good health and prosperity for their Family.

And the Wind Changed again and we moved to Tallahassee Florida.

Now we are in Tallahassee at Florida State University where I have been for the longest period in my career.

LESSON: Because of my background in starting things, I was brought in open and grow FSU's tech transfer office. I did so and because the faculty and university recognized it was needed, the office was straightforward to grow. At this same time, I became active in our professional association, AUTM, in an effort to bolster my contact network to help find corporations and the latest deal terms.

LESSON: This outreach has cemented by faith in networks. Some of my best friends are from AUTM. Due to professional schedules and distance, I may only see them once a year and we may not talk routinely, but there is great security and satisfaction in maintaining an active peer network. Some of your Peers are down the hall, many are not. Networking pays off every time. I was elected volunteer President of AUTM in 2006 and as a result travelled the globe 'communicating the value' of university research to make a "Better World" (just a small plug for our annual publication). I think I'm on my fourth passport and it has definitely made me more of a global thinker.

And the wind changed as I realized there is whole world out there of people just like me -- raising a family and working to make things better. They may speak different languages, but they are just like me.

You learn that if you can improve the economy is developing countries, the Mothers in the country see a Better World for their kids and the family is less drawn to radicalism and the security of their country and our country is more secure.

What does all this have to do with Career Advancement – EVERYTHING. Your own future is in your own hands. The future is yours – Go out and Seize it !

> **Tallahassee, Florida**
>
> - Asst VP Research & Economic Development
> - Movies: Iron Man III, Star Trek into Darkness
>
>
>
> - LESSON: It is very possible that you will be doing something in 10 years that you never thought you'd consider or that doesn't even exist yet. Keep your eyes and ears open always and enjoy the ride!

And the Wind is changing again now as I begin to consider retirement. We are now empty nesters. Our kids are well disbursed across North America and we are fortunate to be included in their family life as they raise our grand kids, of which we have three. It is very satisfying to realize that we have raised three independent "kids" and to now see them respected in their own careers in Hollywood, DC and Vancouver with families of their own.

LESSON: Raising a family is time consuming, but they all leave home and if you don't acquire a hobby where you are passionately involved, you hesitate about retiring. I have no hobbies, so I am expanding my part time consulting activities in place of a hobby. I have a National and International reputation for expertise in university start-up companies, building a workable innovation ecosystem and in getting results. I consult in the US, Canada, Chile and Tunisia. I am always looking for new opportunities. If you know of any, let me know.

LESSON on Career Advancement: By taking full advantage of every opportunity presented to me, both the large and small, the successful and the not-so, I have had a very eclectic career moving from the private sector to government to universities and back. Never underestimate the value of a good degree, but you need to always have your personal radar set for interesting opportunities. Such an opportunistic approach may not make you wealthy, but it's never dull, you'll have stories to tell and you can make a difference.

In closing, for Career Advancement Opportunities, keep in mind that it is very possible that you will be doing something in 10 years that you never thought you'd consider or that doesn't even exist yet. Keep your eyes and ears open always and enjoy the ride!

Chapter 10

A Chemist Strategizing for Chemists: My Career in Associations and Societies

Robert Rich*

Director, Strategy Development, American Chemical Society, 1155 Sixteenth Street, NW, Washington, DC, 20036
*E-mail: r_rich@acs.org

There are many opportunities for those with a background in chemistry to have a rewarding and productive career, while at the same time making a difference through scientific societies and other professional associations. This paper will discuss my personal experience, and provide generalizable lessons which can be applied by anyone considering this career pathway, especially when your educational background aligns with the mission of the society.

Preparing for a Career in Associations

There is no standard curriculum or professional education which prepares one to work in an association. This is an advantage for those scientists, such as myself, who are drawn to the idea of working for a mission-driven organization. The problem-solving, communication, and organizational skills that are typically taught in a scientific education are a good match for such a career.

© 2014 American Chemical Society

My own transition to association leadership started with the Ph.D. program in Organic Chemistry at the University of California, Berkeley. I studied the mechanisms of enzymes which plants and bacteria use to synthesize aromatic amino acids. Under the supervision of Professor Paul A. Bartlett, I was able to design, synthesize, and test a novel probe. I (both literally and figuratively) carved my initials in the door of the accomplished scientists making a contribution in the laboratory (Figure 1).

At the same time, while at Berkeley as a graduate student, I was able to explore the opportunities to get involved in student volunteer activities. Perhaps as a harbinger of things to come, I immersed myself in student leadership, first through the Berkeley Hillel Foundation, rising from a volunteer group leader to joining the board and becoming President. I founded a student newspaper and facilitated a day-long conference for my peers. Eventually, I ended up serving as a trustee for a major philanthropic foundation going far beyond the campus, through this work (Figure 2).

I believe that this unusual (though certainly not unprecedented) involvement in volunteer organizations during graduate school was a good indicator that I had the motivation and ability to succeed in association work. This kind of excitement about volunteering and serving a mission in the community may suggest exploration of career options such as my own.

As I approached completion of my dissertation, I wasn't sure what I wanted to do. Up until that point, it had been fairly straightforward to choose a next step in my career. Now, I began to seek out something that involved working with science and also working with other scientists.

Few have the goal of going into association management while in school, and this was certainly not the case with me. I considered and learned about the available "non-traditional" career options through presentations, articles, and networking events (*1*). These included science journalism, patent law, and government administration; among which I settled upon science policy as the most attractive. I liked (and still do like) the idea of making a difference at the "wholesale" level with science versus individual laboratory supervision.

At first, I applied for a postdoctoral fellowship to study chemical weapons non-proliferation, and this led to participation in the International Conference on Science and Security, held among scientists from China, Russia, India, Pakistan, and the U.S. in Kiev, Ukraine. (See Figure 3) I was inspired by the experience, and began applying a position where I could get my start in the science policy profession.

This took some time and required some creativity. There are a number of Science & Public Policy Fellowships (see below), which are very competitive and I did not receive. I was, however, able to take a postdoctoral fellowship that allowed me to bridge the gap and get experience in science policy work. In general, a postdoc can be a great stepping stone or a waste of time, depending upon how carefully one plans to maximize the value of the experience (See Figure 4). Good advice for maximizing a postdoc's value has been published (*2*). In my case, it was also valuable that the fellowship allowed me to pay the rent while having the flexibility to pursue an upaid internship in science policy one day each week.

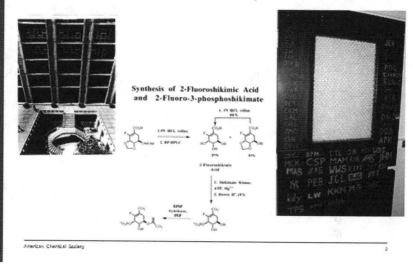

Figure 1. University of California, Berkeley. Courtesy of the American Chemical Society.

Figure 2. Extracurricular activities at Berkeley. Courtesy of the American Chemical Society.

Figure 3. How to get into science policy work? Courtesy of the American Chemical Society.

The Postdoc Experience

- Seven Steps to a Successful Postdoctoral Experience*
    1. Decide <u>where</u> you are going.
    2. To accomplish your long-term goals, set short-term <u>objectives</u>.
    3. <u>Choose</u> the right project (to match your goals, interests and research skills).
    4. Work hard - but most importantly, <u>work effectively</u>.
    5. Set short-term, <u>clear objectives</u> with your advisor.
    6. <u>Broaden</u> your knowledge and experience base.
    7. Keep your eye on <u>your future</u>.
- Also
    1. Pay the rent.
    2. Enable yourself to intern (outside the lab) one day each week.
    3. Be flexible and select a supportive advisor.

*From: http://postdoc.unl.edu/documents/Seven_Steps.pdf

Figure 4. Guidelines for the Postdoctoral Experience. Courtesy of the American Chemical Society.

My postdoctoral fellowship took place in the Laboratory of Bio-organic Chemistry in the National Institute of Diabetes, Digestive & Kidney Disorders at the National Institutes of Health (NIH) in Bethesda. This was an ideal base to be close enough to Washington, DC to participate in the science policy community gathered there. My research was still in bio-organic fluorine chemistry, which I worked on for only five days each week thanks to a flexible supervisor. In this manner, I was able to support myself while getting valuable experience in my one-day-a-week internship. I was able to make a connection through my alumni network with the Director of Science & Policy Programs at the American Association for the Advancement of Science (AAAS), who allowed me to provide staff support to their Task Force on Careers for Young Scientists. Getting a job in associations requires both relationships (i.e., who you know) and experience (i.e, what you know), and this internships provided me with both. I also was able to take part in a three-month rotation in the NIH Office of the Director, which gave me a perspective on strategic planning & evaluation (Figure 5).

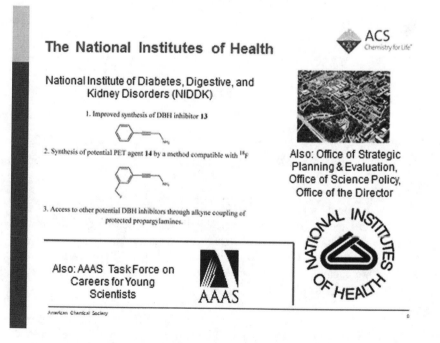

Figure 5. *The Year of Postdoc at NIH/Unpaid AAAS Internship. Courtesy of the American Chemical Society.*

Working in Associations

There are many different pathways within the association management profession. Some of these are listed in Figure 6.

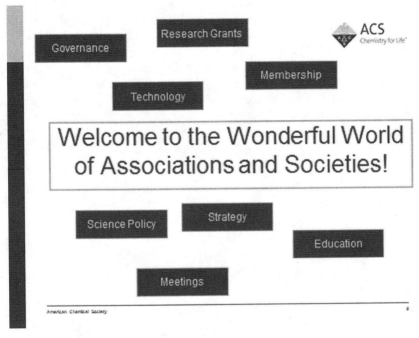

Figure 6. Some Jobs in Association Management. Courtesy of the American Chemical Society.

In general, the skills required include the following:

- **Big picture thinking**: Being able to see the needs of the entire profession as well as those of each individual practitioner is important. This kind of thinking is sometimes at odds with the intense focus required to do cutting-edge research.
- **Problem solving**: The interactions of many members, leaders, volunteers, and staff can create challenging systems problems, for which an analytical scientific background is good preparation.
- **Creativity**: Often, association leaders need to create innovative approaches to be most successful. This is often a hallmark of those educated in science.
- **Written and oral communication**: A big part of association work involves writing emails, reports, and other documents. It is also essential to be able to present your thinking at meetings and in conversation. Not all scientists are good at this upon graduation, and may benefit from an unbiased appraisal of their ability to communicate. Fortunately, this is a skill that can be improved with practice.

- **Desire to work with other people**: Association management involves constant engagement with members, colleagues, and others. Again, motivation and the skills to do this are not always the hallmark of scientists.
- **Mission-driven vs. profit-driven**: While maintaining financial sustainability is important in the not-for-profit sector (which includes almost all associations), finances are secondary to the organization's mission. This implies that salaries in this sector aren't as high as they can be in for-profit companies, and that a part of the satisfaction of these jobs is the ability to make a difference in improving people's lives. It is very helpful to be enthusaistic about an association's mission in order to be happy working there.

Preparation for a career in this sector should emphasize getting experience in volunteer organizations as a student, practicing written and oral communication, and developing a network of contacts within the association community.

Science Policy

The first major area of association management that I participated in was science policy. Just about all scientific societies and associations, whether 501c(3) or 501c(6), have a science policy, public affairs, or related department. This work focuses on conveying the concerns of scientists to policy makers, helping to set the parameters under which science is funded and conducted, and applying scientific findings to societal needs.

The key success factors for working in this area include developing a good understanding of the legislative process and how government operates, and having a strong network of people well-connected in the legislative and executive branches. It also is valuable to cultivate ties with leaders of the scientific community, including those from academia, industry, and government laboratories.

The most visible mechanisms to enter the world of association science policy are the science & technology fellowship programs sponsored by AAAS, the American Chemical Society (ACS), and other associations. These highly competitive programs include a comprehensive training/orientation program and placement within a Congressional staff office, an Executive Branch agency, or an association science policy department. As my own experience demonstrates, however, this is not the only way in. Creative approaches to getting the requisite experience and expanding one's networks are possible.

Following my unpaid internship (and simultaneous postdoc/policy rotation at NIH), I was offered a full-time job to help start the AAAS Research Competitiveness Program. This program provides expert advice to those investing in, supporting, and managing activities in science and technology around the world. It assembles and leads carefully tailored teams of scientists, engineers, policy makers, and innovators to provide expert peer review and guidance to research institutions in such areas as technology-based economic development, strategic planning, capacity building, and evaluation. It draws upon a network of experts that includes researchers from all areas of science and technology, persons experienced in product development and technology transfer, and entrepreneurs and business managers from all fields. It is administered by, and draws upon the resources of the AAAS Directorate of Science & Policy Programs.

Career and Professional Development

After spending several years at AAAS, I got a phone call from a colleague at the ACS with whom I had worked during my internship year. She was also familiar with my work in an ACS member volunteer capacity during my undergraduate and graduate school education. The ACS was looking for someone to lead the Office of Professional Services in the Department of Career Services. My network had once again provided a key opportunity for career advancement.

Career and professional development is a key activity in most professional societies and associations. In order to be successful in this work, it is important to enjoy helping other people to have successful careers, and to be a good mentor and advisor. Aspects of career development include running educational workshops, one-on-one career consulting, local programs to provide networking and advice, job fairs (face-to-face and virtual), and online resources (e.g., guides to interviewing or resume writing skills).

Research Grants Administration

After working in professional development, I became a Program Officer within the Research Grants Office, helping to administer the ACS Petroleum Research Fund. While not all associations have quite the same size of research foundation (PRF at the time gave out $25 million per year in support for academic research.), most do have some external grants that are administered.

The key skills for success in grants administration include familiarity with the field for which grants are provided, being able to effectively communicate with the applicants, and being facile with accessing the information required to review and select the winning awards. In my position, I was required to identify the best peer reviewers for a large number of proposals, quickly understand the focus of diverse research projects, and stay connected with the research community.

This job is similar to those of program directors at other governmental (e.g., NIH, NSF) and non-governmental (e.g., Dreyfus, Welch Foundations, HHMI) research funders. Most people, including myself, get these jobs through a combination of research experience and network connections to those already working in these agencies.

Strategy Development

For the last five years, I've been working in the development of organization-wide strategy and planning for the American Chemical Society (see Figure 7). This kind of position, rare within small or mid-size associations, is common in the largest organizations (e.g., ACS, IEEE, National Geographic), but often with differing job titles and detailed responsibilities.

At ACS, the Strategy Development team supports ACS volunteer and staff leaders to:

1. Develop and communicate Society strategy, including the ACS Strategic Plan,
2. Construct, and encourage collaboration around, strategic initiatives such as environmental scanning (i.e., futures research), worldwide strategy, and Presidential initiatives, and
3. Engage with governance (volunteer leadership) bodies and cross-cutting staff groups to enhance strategic alignment.

Thomas Jefferson said "an association of men who will not quarrel with one another is a thing which has never yet existed, from the greatest confederacy of nations down to a town meeting or a vestry." Every association needs someone to help facilitate this "quarreling," so as to channel it into productive avenues, developing synergies across organizational barriers. In smaller groups, this function is a responsibility of the Chief Executive Officer or Chief Operating Officer. In larger organizations, dedicated staff can be assigned to help develop constructive and aligned strategic management.

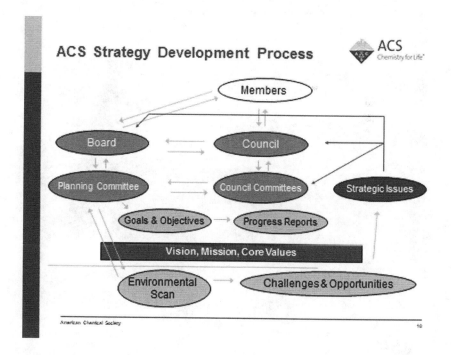

Figure 7. The ACS Strategy Development Process. Courtesy of the American Chemical Society.

At ACS and most other associations and societies, strategy ultimately derives from the professional needs of individual members. Their representatives, elected to the ACS Board (16 members) and Council (about 500 members), work through committees to establish Goals and Objectives in support of the overarching Mission, Vision, and Core Values. Progress is reported periodically against these Objectives. All of this planning rests upon an understanding of the Society and external factors developed through "environmental scan" research which leads to the identification of important challenges and opportunities. The research also identifies key strategic issues requiring in-depth deliberation. The Strategy Development team facilitates all of this planning, research, and deliberation by preparing documents, framing conversations, and sharing their results with others within the organization.

The culmination of the planning effort results in the public *ACS Strategic Plan*, which is posted on the Web at http://strategy.acs.org (Figure 8). This dissemination, supported by the production of trifold pocket card summaries, enables a broad range of ACS members and staff to know of the plan's Goals and to work together to achieve them.

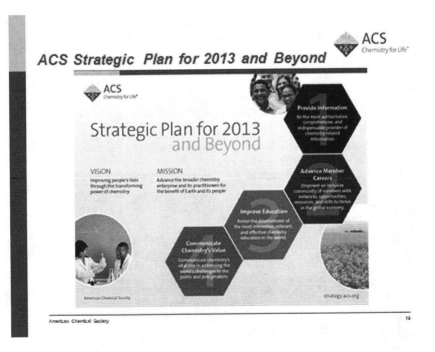

Figure 8. The ACS Strategic Plan. Courtesy of the American Chemical Society.

Some recent Strategy Development activities beyond strategic plan development are listed in Figure 9.

Some Interesting Projects in ACS Strategy Development

- Environmental Scanning
  - Three Horizons Model for Ongoing Dialogue about External Trends
- Scenario Planning
- Sustainability Engagement Event (SEE)
  - Appreciative Inquiry Summit
- ACS Worldwide Strategy
- Technology Strategy Roadmap
- Governance Review
- Promoting an Innovative & Collaborative Culture

Figure 9. Some Interesting Projects in ACS Strategy Development. Courtesy of the American Chemical Society.

Work in strategy development requires integrative and systems-thinking skills to be able to see the big picture of the organization. Diplomacy and interpersonal excellence are required to be able to build consensus around developing strategys and facilitate decision-making among competing priorities. Since a deep understanding of the organization and its complexities is essential competency, strategy is not usually an entry-level job. Most often, people are hired based upon internal connections and experience. In order to support the crafting of the most effective strategies, a broad knowledge in a variety of areas is essential, including:

- Associations
- Strategy
- Science
- Volunteer Leadership
- Technology
- Future Forecasting
- Executive Management
- Membership, Publications, Scientific Information, Education, Research, and other components of the organization

Some Other Staff with Science Degrees at ACS

- CEO and Executive Director
- Director, Education
- Director, Research Grants
- Director, Membership & Scientific Advancement
- Director, International Activities
- Director, Green Chemistry Institute
- Manager, Innovation & Collaboration
- Vice President, Journals Publishing Group
- Vice President, Editorial Office Operations
- Assistant Director, Career Management & Development
- Editor, Chemical & Engineering News

Figure 10. Some Other ACS Staff with Science Degrees. Courtesy of the American Chemical Society.

Associations as a Profession

While few pursue a degree in science with a career in associations and societies as the ultimate objective, such work can provide excellent opportunities to improve people's lives. The profession of association management is well-established, with substantial opportunities to share best practices, network, and collaborate with other organizations.

The largest association industry group is the American Society of Association Executives: The Center for Association Leadership, commonly referred to as ASAE. This group of more than 21,000 professionals, in all aspects of association work, gets together annually for a large conference, and holds many focused meetings, conferences, seminars, and events year-round. ASAE maintains an online career center, which can be used to search for job opportunities in associations nationwide.

For those working in scientific and engineering societies, the Council of Engineering and Scientific Society Executives (CESSE) has an annual conference and networking resources. There are currently 198 member organizations of CESSE, each with a number of active professionals (*3*). Digital Now, organized by Fusion Productions, is another association industry conference, specializing in technology and strategy. There is an informal Young Association Professionals (YAP) organization, which works primarily through social media to connect those starting out in the field.

There are many job opportunities for scientists to consider at associations. Often, positions are advertised on the Careers section of the organization's Web site. As with any profession, however, there are far more opportunities accessible through networking with those currently employed in the sector. At ACS alone, many executives, managers, and staff come from a scientific background. A sampling of these is listed in Figure 10.

Conclusions

It does take a special mindset to excel in associations and societies. A strong belief in the power of people, working together, to effect change is very helpful. Margaret Mead famously said "Never doubt that a small group of thoughtful, committed citizens can change the world: Indeed it's the only thing that ever has." Associations improve people's lives through such collaborative effort. While this kind of work is probably not what many have in mind when pursuing a degree in science, the value that such a degree can provide in this industry is substantial. If you want to make a difference and like working with people, association work could be for you.

References

1. Resources for career management and identification of possible career pathways can be found at www.acs.org/careers and www.aaas.org/careers.
2. See http://chronicle.com/article/Making-the-Most-of-Your/66265/ or http://postdoc.unl.edu/documents/Seven_Steps.pdf.
3. See www.cesse.org.

Chapter 11

From Medical School Dropout to Editor-in-Chief of C&EN

Rudy M. Baum*

Editor-at-Large, *Chemical & Engineering News*, 3222 Miller Heights Road, Oakton, Virginia 22124
*E-mail: rmbaum@cox.net

At the end of the arc of a career such as mine, much seems inevitable. My path from leaving medical school after one year to becoming editor-in-chief of *Chemical & Engineering News* was hardly a straightforward one, however. The theme that underpins my career has been an undying love of language and using it to communicate science. In today's economic climate, it is hard to recommend that one pursue a science writing career. However, I can heartily recommend that young scientists be as passionate about being able to communicate their science as they are about doing it.

I retired as Editor-in-Chief of *Chemical & Engineering News* (C&EN) in September 2012. In my remarks at my retirement party, I pointed out that, at the end of one's career, the events that shape a career often look inevitable. They are anything but, however.

How I became Editor-in-Chief of C&EN illustrates the role of chance, the importance of mentors, and the value of flexibility in the arc of a successful career. It also shows why a chemistry degree opens doors far beyond the laboratory.

I attended Duke University from 1971–1975. I was a pre-med, but Duke did not grant a degree in pre-med. The university insisted, wisely I think, that students had to choose a major in a specific subject. I intended to major in biology, but a fantastic advanced introductory chemistry course taught by Prof. Al Crumbliss convinced me to major in chemistry.

I received a B.A. degree in chemistry. There was a real difference between a B.A. and a B.S. in chemistry from Duke at that time. The B.S. degree resulted

© 2014 American Chemical Society

from the ACS approved curriculum; the B.A. degree curriculum was slightly less rigorous. The main difference that I remember is that the B.S. required a two-semester course in physical chemistry, while the B.A. required only a one-semester course. Because I was a pre-med and interested in taking quite a few non-science courses, I chose to pursue the B.A. degree.

I spent one year at Georgetown Medical School in Washington, D.C. I was a good student, but for many reasons, I had decided that I didn't really want to become a doctor. I took a leave of absence from Georgetown, never really intending to go back. I was pretty certain that I would go to graduate school to get a Ph.D. in chemistry, but I needed a break from academe.

Of course, I needed a job. In those days, looking for a job involved looking at classified advertisements in newspapers. I responded to a "blind ad" in the *Washington Post*. A blind ad told you what an employer was looking for, but didn't identify the employer. It might look something like this:

CHEMIST
National scientific organization seeks bachelor's degree chemist
for staff associate position at its headquarters in Washington, D.C.
Send résumé and cover letter to:
    Washington Post Classified Advertisements
    P.O. Box 3241
    Washington, D.C. 20036

The ACS Personnel Office called me a few days later and set up two interviews in the Education Department. I was actually offered two staff associate positions, and I took one in the Office of Continuing Education coordinating the Society's Short Courses. I did that for a little over a year, and then moved to become the editor of what were called Audio Courses. It seems pretty primitive today, but they were courses on very technical topics and consisted of lectures on cassette tapes coordinated to a printed manual.

After three and a half years in the Education Department, I was ready to move on. I had a vague notion of wanting to be a writer, but the short stories I had been writing in my spare time had not sold. I was half-heartedly looking into chemistry graduate degree programs. Then one Friday in early November 1979 I was procrastinating at work and looked at the bulletin board in the eight floor hallway. We actually had bulletin boards then, and notices and job postings on three-by-five cards were pinned to the cork board. There was a card that looked something like this:

CHEMICAL & ENGINEERING NEWS
  Assistant Editor
  Washington News Bureau
Position is responsible for reporting news and events relevant to
chemistry, the chemical industry, and the American Chemical
Society occurring in the federal government.
Bachelor's degree in chemistry required; reporting/editing
experience desirable.

I immediately called the Personnel Office and indicated that I would like to apply for the job. It turns out that the job posting at C&EN had been up for the required two weeks and was going to be removed at the end of the day. The following week I interviewed at C&EN, but the interview was odd; I talked to two senior staff members who were involved in editing and production at C&EN, not reporting on the federal government. I found out that the assistant editor who worked in editing and production had applied for the government reporting position. The senior team at C&EN wanted to give him the job, but they didn't know who they would replace him with. That turned out to be me.

It gets stranger. I started at C&EN in January 1980. Sometime during the spring, the magazine's West Coast Bureau Head resigned from C&EN to take a reporting position at *Science*. My hope was that one of the science and technology reporters in Washington would want to take the plum position in San Francisco and open up a writing job for me in Washington. I didn't have any hope that C&EN's Editor-in-chief, Mike Heylin, would send a completely inexperienced reporter to a one-person field office in California.

Months rolled by. It was a time of very high inflation—16% or more per year. The economy was on the verge of a pretty severe recession. People were reluctant to pull up stakes and relocate. People were reluctant to change jobs. No one applied for what in the past had been one of the most sought-after jobs at C&EN. One Friday afternoon in November, the magazine's staff editor, Ernie Carpenter, who I sort of reported to, came into my office as he was leaving for the day and asked me if I had ever thought about applying for the West Coast position. I said I had, but I didn't think Mike Heylin would send me because I had no experience as a reporter. He said something like, "He probably wouldn't have in the past, but he's so disgusted with the situation, he just might consider you."

Talk about a ringing recommendation! On Monday, I officially applied for the job, Mike interviewed me, and then ... nothing. Two weeks went by. It was driving me nuts. Then, once again on a Friday, Mike called me into his office and told me that I was going to be C&EN's reporter in California.

I became C&EN's West Coast Bureau Head in March 1981 and spent 14 wonderful years covering chemistry in the western half of the U.S. and a variety of topics that interested me. I introduced coverage of the AIDS epidemic into C&EN. I covered the discovery of C_{60} and the fullerenes. And I met an incredible number of outstanding chemists at University of California, Berkeley, Stanford, Caltech, UCLA, and many other universities.

In 1993, Mike Heylin decided that, for a number of reasons, C&EN needed a true managing editor, a position that had not been filled at the magazine for more than a decade. I was contacted by the executive search firm hired to fill the position and told that I was an internal candidate for the position if I was interested. I said I was. The interview process lasted several months and resulted in Madeleine Jacobs being hired. That fall, before C&EN's annual Advisory Board/ Staff meeting, Madeleine told me she would like me to return to Washington to become the magazine's assistant managing editor for science, technology, and education coverage. I started in that position in August 1994.

In July 1995, Madeleine succeeded Mike as C&EN's Editor-in-chief, and she named me managing editor. In 2003, Madeleine applied for and was selected to succeed John Crum as Executive Director and CEO of ACS, and I succeeded Madeleine as Editor-in-chief of C&EN.

See what I mean? Med school dropout. Blind ad in the *Post*. A chance glance at a bulletin board I never looked at. Bad economic times. Self-taught journalist. All of it adding up to a very successful career.

C&EN was a great fit for me. I loved chemistry, but I was never a lab rat. I loved the intellectual side of chemistry much more than the practical side. In my career, many brilliant chemists around the world were my lifelong teachers. I was basically a graduate student for life who didn't need to spend endless hours in the lab. Interestingly, when I started at C&EN, I thought I would eventually move on to a more general publication, but I found that I liked writing about chemistry at a level geared toward informing other chemists about developments in our science. I never had to define a "chemical bond" in my stories.

So what are the career opportunities for science writers? I wish I could say they are bright, but I can't. Journalism has been in a state of crisis for the past decade or more. The internet has been a truly disruptive technology for journalism because it has upset journalism's longstanding business model. More people are reading newspapers and magazines than ever before, but for the most part they aren't paying for the content they are consuming. Advertisers have deserted print publications in droves. Science journalism has taken a particular beating. Many newspapers, magazines, and television and radio stations dropped science reporting altogether.

That National Association of Science Writers (NASW) was founded in 1934. The organization currently has 2,254 full members and 269 student members. I went through the member roster—this was not a rigorous analysis!—and found 156 members working for news outlets. I found only 22 publications that employed two or more NASW members. Some of them are listed in Table 1.

Table 1. Publications employing two or more NASW members in 2014

Science	17	Popular Science	3
C&EN	14	American Scientist	2
Nature	13	Discovery News	2
Science News	10	EARTH	2
Scientific American	9	IEEE Spectrum	2
New Scientist	7	Los Angeles Times	2
Discover	6	Sky & Telescope	2
NPR	6	Technology Review	2

Some of these are venerable science magazines, but it is significant that the top three are publications that are primarily written for professional scientists.

There are only four newspapers, one radio station—really a radio network—and no television stations.

The good news is that there are more than 2,000 other NASW members. About half of these NASW members are freelancers and about half work as public information resources at universities and nonprofits like ACS. In fact, ACS does employ several NASW members in addition to those who work at C&EN. These are very talented science writers who are playing a vital role in keeping the public informed of developments in science and technology.

So, are there career opportunities for communicating science? It is not a huge field, but it exists and it is a wonderful way to be involved in the chemistry enterprise even if you do not want to be a practicing chemist.

Chapter 12

Consulting for Seniors: Having It Your Way

Thomas R. Beattie*

Independent Consultant, 2648 Angell Avenue, San Diego,
California 92122-2103
*E-mail: beattietr@aol.com

Consulting can provide professional satisfaction and some personal income for senior/retired chemists. The author reviews, based on his own consulting experiences, the many factors that lead to a successful consulting business using accumulated personal knowledge and experience. Consulting may not work for everyone. Although it is difficult to do full-time, there are many avenues to explore in looking for part-time consulting. If you try it, it can be invigorating and stimulating. If you try it and it is not working well for you, get rid of it quickly and find other things to enjoy.

Introduction

An individual's ideas about "consulting", like ideas about "motherhood" or "government", are very much dependent on that person's background, outlook and experiences. Based on his own consulting experiences, the author has found consulting may represent an opportunity worth developing.

Now, almost anybody can be a consultant. I have met young ones (fresh out of school without a job and therefore a "consultant"), a recently-let-go computational chemist who decides to work part time and equal his former income, and, among others, an unemployed chemist who anticipates keeping a roster of ~20 companies for consulting. Indeed, for $50 one can buy 500 business cards and become a consultant.

Being a SUCCESSFUL consultant requires some attributes which make it an ideal choice for senior chemists to consider. Resourcefulness and wanting to make things happen; having had a long-term, varied, successful career; being excited by a professional challenge rather than money; flexibility and adaptability; and being professionally current are all attributes leading to success as a consultant.

© 2014 American Chemical Society

Avenues To Explore

For example, a chemist in retirement from a career in a single big pharma company probably has multiple areas of expertise: synthesis, program development, working at the interface of chemistry and biology, data management, computational chemistry and patent crafting are some. Also, that person probably has people management and functional management skills and perhaps even project management and commercialization and marketing skills.

More the norm these days perhaps, a chemist with a working career in several different companies can use the varied experiences from how different companies handle issues and problems to advantage in consulting.

All of these skills are useful in big pharma, biotech and academia, although the operating styles and needs of each are very different. The trick, of course, is to find a niche which is in need of the skills you possess.

Big pharma, in a period of rapid change for at least the last decade, may not be a current fertile hunting ground for consultant spots. Most often they are rich in talented chemists and basically have been in static or retrenchment mode. Clearly they have lots of experts, perhaps too many in the corporate view.

Biotech is diverse in its needs because the diversity of biotechs in number of people and size of financial assets is so large. Ranging in size from one to thousands of people and assets from almost no money to billions, they offer a huge opportunity for chemists to connect in helpful and meaningful ways. A company of a handful (or fewer) certainly does not contain all the knowledge and skills needed to find and move along a drug candidate or a medical device. Since there are so many, it is difficult to know who is in need of the skill set you possess.

Academia offers different kinds of opportunities. The chance to fulfill a wish to teach, the opportunity to provide support within a research group, and the mentoring of developing young scientists are all likely functions for participation of senior chemists.

Thinking further afield, I can suggest looking into law firms, which may need technical assistance with patent applications and protection or with expert witnessing. With the rise of philanthropy in support of research, angel investors, philanthropic foundations and venture capitalist groups each may have need of experienced senior chemists.

While the range of possible opportunities is large, the problem of connecting with the right situation is large as well. However, once the connection is made and consulting is underway, the rewards can be satisfying. In my own situation, I have had consulting assignments in most of the areas mentioned above, and one aspect very satisfying to me has been the diversity of the things I was asked to do.

Some Differences

Consulting is different from full-time employment in that it is usually not full time. That encompasses many differences. Compensation is a major one. Your compensation is not set – you negotiate it and it usually doesn't come regularly every second Friday. It may be based on per hour or per task or it may be zero, if you agree. If you are receiving social security benefits, your consultant income may affect your monthly benefit payment, depending on your age. Payment may not be timely. No taxes are usually withheld, possibly making it necessary for you to file quarterly estimated tax payments.

A great benefit: since it qualifies as "earned income" you may have the opportunity to use some/all of your compensation for making Roth IRA contributions. If you are a joint income tax filer, your joint partner can be included in your Roth contribution calculations, thus possibly doubling what you invest together.

While the above paragraph focuses on compensation, it probably should not be the major motivation in deciding to consult. It is difficult to match a full-time position annual salary, and to do so you will probably need to work just as hard and long as if you were full-time employed. If that is your goal, you might just as well work full-time.

The consulting I have in mind ("Having it your way") allows for flexibility and adaptability – you are doing it on your terms. It incorporates fitting the consulting tasks to your current/retired life style.

You control the timing - do you want long or short term assignments, how intensely do you want to work, do you want part or full days of work, do you want to block out time for travel and periods of not consulting?

If you consult locally, it may be possible to do it at times so as to avoid rush hour traffic. Also, depending on the work type, work can often be done at home, thus avoiding travel completely.

In contrast, assignments out of town may afford the opportunity to include personal time before or after consulting visits. You save on airfare, which the client is paying, and sometimes you get to visit some interesting and delightful places.

How To Start?

No surprise here – networking and personal referrals. Networking is how I and most others get consulting assignments. I chose to keep things simple, so I don't advertise. If I receive an inquiry from someone to whom my name has been given about providing help, I have already been at least partially validated through the professional referral. That is critically important.

If you do want to reach out for business, there are many avenues to explore. Your local ACS section and other professional groups have programs that you can attend and gain visibility. The ACS website, www.acs.org, has guidance to get you started.

Remember, making the connection is difficult and requires diligence, patience and a bit of luck.

Some Pros and Cons

Don't get me wrong, I like money too. But I find the best reason for seniors to consult is the professional challenge: the opportunity to make an impact, the respect you acquire when you make a suggestion no one else thought of, and the chance to prevent someone from going down a pathway you know to be unproductive. You will be meeting new faces and can be buoyed by their intensity and enthusiasm.

I especially enjoy the opportunity to have an impact on younger scientists and to watch them grow. Working with younger scientists is a good way to keep yourself feeling young.

There are some downsides, of course. You should be providing an unbiased opinion and sometimes your clients won't appreciate what you are telling them. That situation can be tough to take and not easily reconcilable. Also, things change and sometimes urgency pops up in unanticipated ways. A deadline gets moved up, requiring more immediate work time and effort than you anticipated. Project themes and goals may change in ways inconvenient for you. A hurried call asking you to drop everything and come now sometimes happens too. "I need the answer now" happens in consulting just as in full-time employment. And, in working with multiple clients, somehow they often all seem to need the final product of your effort at the same time.

Some Examples

In my 14 years of consulting I have worked on a wide variety of projects. Some were very specific and of limited duration: helping a small company vice president (with a biological background) hire a new director of chemistry in which my recruiting, resume reading, and interviewing skills were important; visiting several different small companies and passing judgment on which compounds looked most promising to pursue and which compounds might be problematic to follow up; evaluating an issued patent and providing guidance to attorneys on exactly what the patent meant in the time frame of 20 years ago.

Some were much longer term: helping in a very successful chemistry tools company to spin off a chemistry department into a sister company. In this case the initial contract was for "up to 40 hr. weekly for up to 6 months duration", which at the time sounded very much to me like full-time employment. However, I found the people, the challenge of getting it right, the chance to create a new organization, etc. so invigorating that I transitioned into stages of helping the architects design new lab space, preparing the labs for move-in, staffing the new lab space with newly hired chemists, managing the operation of the labs, participating in business development to find new customers, etc. that the "up to 6 months" morphed into almost 2½ years of almost full-time work. And I let it happen that way because I was enjoying it so much!

And then there were also jobs of intermediate duration: participating in weekly chemistry meetings and biweekly project team meetings at a small anti-cancer company over a period of ~1 year; a 3 month tour-de-force of working on surveying the patent literature in a company which already had launched a product and was evaluating related research areas to see which looked promising for entry; a nine month stint in a company in which I did all kinds of chemistry things that the primarily analytical chemistry staff was not equipped to do. For a major pharma company I worked off and on with 3 separate divisions doing third party due diligence inquiries with academics and small companies.

And there were other assignments too. Hopefully, the examples above have made my point. You can use your skills and knowledge base to jump in where needed and have an enjoyable experience while providing much-needed help.

One sort of opportunity I haven't chosen to accept: there are many individual entrepreneurs and very small (1-5 people) companies where the operating funds are just not yet available. Opportunities abound to join and work for the promise if things go as we hope, then we will have the money to pay. I did that once, and did not find it satisfying.

Business Issues

If you consult, you are a businessperson running a small company. Your locale may require a business license. You need to decide whether to operate as a sole proprietorship or incorporate as a company. Do you plan to have a home office? Are you planning to advertise your availability to consult? You will pay taxes on your business income. For social security purposes you are an employer and an employee, so you pay double – generally twice the rate you paid as a full-time employee in someone else's company. You may wish to have insurance coverage. You may need legal help with some of these issues.

I want to stress again that for most chemists consulting should not be looked upon as an alternative to full-time employment. As I have found, it's an up and down thing – sometimes you are busy and other times you are not. Further, my interests have evolved over time as to how much I chose to do with consulting. For me, that is the fun of it – I am having it my way.

In summary, consulting may/may not work for you. It is difficult to do full-time, but there are many avenues to explore in looking for part-time consulting. If you try it, it can be invigorating and stimulating. If you try it and it is not working well for you, get rid of it quickly and find other things to enjoy.

Innovation and Entrepreneurship

Chapter 13

The American Chemical Society (ACS) Entrepreneurial Initiative

Roger E. Brown,[1] Elizabeth I. Fraser,[1] Kenneth J. Polk,[2] and David E. Harwell[*,1]

[1]Department of Career Management and Development, American Chemical Society, 1155 16th Street NW, Washington, DC, 20036, United States
[2]Office of the Executive Director and CEO, American Chemical Society, 1155 16th Street NW, Washington, DC, 20036, United States
*E-mail: d_harwell@acs.org

The ACS Entrepreneurial Initiative is a members-only program with a broad reach into the entrepreneurial support community. ACS members and chemical entrepreneurs alike benefitted from ACS's pilot experiment, which was conducted during the 2011 to 2013 time period. From the pilot, ACS members obtained materials and instruction about entrepreneurship from the Entrepreneurial Training Program and access to information, professional services, mentors, and sources of funding or partnering for their start-ups from the Entrepreneurial Resources Center. Program participants have stated that the pilot offerings made a big difference, and their collective voices have told us that ACS can and must do more. Program participants have in turn stimulated the broader entrepreneurial community, and the assets produced through the program are being more widely distributed to colleges and universities through partnerships with campus innovation and entrepreneurial centers.

Introduction

Since 2008, more than 25,000 jobs—including thousands in research and development (R&D) (*1*) —have been lost in chemical manufacturing companies in the United States, and layoffs continue. Between 1989 and 2009, a clear job loss trend is evident in Bureau of Labor Statistics data that suggests the loss of

© 2014 American Chemical Society

approximately 300,000 full-time chemist jobs in the U.S. (2) Patterns of hiring are also changing. Chemical companies with more than 500 employees are hiring significantly fewer new graduates than in the past, while small businesses are hiring more, albeit at slower rates. While no single factor explains these recent job losses or trends, higher input costs, shrinking margins of large companies, and growing aversion to the risks and costs of investment in longer-term R&D appear to play significant roles.

ACS Presidential Task Force on Innovation and Job Creation

In early 2010, ACS President Joseph S. Francisco appointed a Presidential Task Force to explore the causes for these historic job losses and to recommend ways that ACS could help stimulate innovation and encourage the creation of jobs across the chemical enterprise. The Task Force was chaired by George Whitesides, the Woodford L. and Ann A. Flowers University Professor at Harvard University. It comprised eminent members of the chemical enterprise from industry, academia, and government, all with experience in entrepreneurship. They included Henry Chesbrough, University of California, Berkeley; Pat N. Confalone, DuPont; Robert H. Grubbs, California Institute of Technology; Charles Kresge, Dow Chemical; Michael Lefenfeld, SiGNa Chemistry; Chad A. Mirkin, Northwestern University; Kathleen M. Schulz, Business Results, Inc.; and Timothy M. Swager, Massachusetts Institute of Technology.

The Task Force issued a report that was published online at www.acs.org/creatingjobs. Recommendations of the Task Force are listed below.

Task Force Recommendations

1. ACS should develop a single organizational unit—a kind of "technological farmers' market"—offering affordable (or free) help to entrepreneurs.
2. ACS should increase its advocacy of policies at the federal and state level to improve the business environment for entrepreneurs and startup companies.
3. ACS should work with academic institutions and other relevant organizations to promote awareness of career pathways and educational opportunities that involve or include entrepreneurship.
4. ACS should increase public awareness of the value of early-stage entrepreneurship in the chemical enterprise with focused media coverage and information targeted to federal agencies that support chemistry.

ACS Entrepreneurial Initiative

In August of 2011, the Society Committee on Budget and Finance recommended funding and the Board of Directors approved a 2-year pilot program known as the Pilot Entrepreneurial Initiative (Pilot EI) to explore training and resources for chemical entrepreneurs through ACS.

The Pilot EI consisted of two parts: the Entrepreneurial Resources Center (ERC) and the Entrepreneurial Training Program (ETP). The ERC was formed in response to Task Force Recommendation 1, and the ETP in response to Recommendation 3.

The ERC is a virtual marketplace of entrepreneurial resources, where free, shared-cost, or reduced-cost access to ACS's scientific information resources (CAS SciFinder and ACS Publications' journals), ACS member expertise, and ACS's key professional service providers (e.g., attorneys, finance, IT, human resources, and marketing) are provided to program participants. Participants can use these tools to plan, create, launch, operate, and grow their chemical start-ups. The Center provides participants with introductions to successful chemical entrepreneur mentors, sources of capital, and larger chemical innovators. The Center's use of ACS's core strengths in this manner provides ERC participants with a unique environment for accelerating the planning, creation, and growth of chemical start-ups.

The mission and goal of the ETP was to provide ACS members with valuable training to advance their interest in pursuing entrepreneurial careers. The program licensed course content from the Kauffman Foundation's FastTrac program. In 2012, ACS granted $500 scholarships to program participants who were selected through a competitive application process. Scholarship winners did not receive payment until they completed FastTrac training through existing course providers. In 2013, ACS brought the training in-house.

Results of Pilot

The ERC accepted 20 start-up companies out of 26 which applied during its first round of competition in July 2012. As conceived, each of the company founders received access to ACS Publications journals, CAS SciFinder, key service providers, advisors, and introductions to sources of capital and larger innovation partners.

ACS also teamed up with the San Diego Entrepreneurial Exchange (SDEE) (a local San Diego network of more than 1,600 chemical and other entrepreneurs) to host a pitching event at the Jansen Facility in La Jolla, CA, on December 11, 2012. This was called the ACS Showcase West (See ACS Showcase West Program, Appendix 4-1). Representatives from 19 companies advancing chemistry-based technologies pitched for funding to an in-person and online audience of angel investors, venture capitalists, and commercial partners. Of the 19 companies, 5 were ERC participants. Those presenting were eligible to receive a one-time $5,000 cash award for the "best pitch". Company pitches were judged by a team of investors and Entrepreneurial Initiative Advisory Board (EIAB) representatives. The winning "best pitch" presenter was a local San Diego area company – Eolas Therapeutics – with a first in class therapeutic for smoking cessation/nicotine addiction. Seven of the companies that presented (two from ACS's ERC) are now in discussions with potential investors as a result of the event.

In January 2013, 11 of 14 start-up companies applying for entrance into the ERC were admitted into a new class for a 6-month term. These companies are currently utilizing the ERC's resources including ACS Publications and CAS SciFinder.

On March 27 and 28, 2013, the ERC successfully held ACS's first-ever Entrepreneur Summit at the Chemical Heritage Foundation in Philadelphia, PA (See ACS Entrepreneurial Summit Program, Appendix 4-2). The Summit was overwhelmingly positively received, with about 93% of those attending being satisfied with the program content and forum. A total of 83 people attended the Summit (51 registered attendees, 26 presenters, and 5 staff).

Two days of programming were included in the Summit which informed participants about pitching to investors, financing strategies, insights into strategic partner processes, challenges to commercializing chemical technologies, methods to demonstrate product value, to name a few topics. Presentations by experienced entrepreneurs and seasoned speakers made this forum a helpful addition to the ERC's offerings. From this forum alone, ERC participants gained far greater insight into the business, legal, marketing, and finance aspects of commercializing products. Presentations from the summit were streamed online to allow off-site viewing.

During the pilot program, the number of applications and the number of participants in both the ETP and ERC were lower than originally anticipated (see pilot EI metrics, Appendix 3). Studies indicate that marketing and awareness campaigns should be ramped up and optimized to capture the potential market for these programs.

In 2012, a total of 56 applications were received for the ETP, and 29 were approved by the EIAB for entry to the program. During this first year of the pilot, approved applicants for the ETP received partial reimbursement for course registration costs through $500 scholarships which were distributed upon completion of training. ACS paid out $5,500 in scholarship funds for the program to 12 people. Although a total of 29 applicants were approved, only 12 people completed face-to-face training and qualified for reimbursement. Sixteen people never started the program, and two started but did not complete the training program. The primary reason cited for people who did not start the program was the burden of paying for the courses up front (tuition ranged from $700 to $1300), and the difference between the actual cost of the course and the amount received through the scholarship. For many, the financial burden was too high.

Based on this feedback, the Pilot EI retooled. Starting in 2013, ETP participants did not incur a financial burden, because the course was offered at no cost to them online through ACS. As stipulated in the original funding request for the Pilot EI, the first year of the program was used to train facilitators for the course, and to build an online infrastructure. These preparations allowed ACS facilitators to offer the course directly to students online in 2013. In January 2013, 38 applicants were accepted into ETP out of the 56 that applied. The eight-week, online course started on February 26 with 39 participants.

Conclusions and Future Directions

The pilot Entrepreneurial Initiative (pilot EI) offerings were based on the recommendations of then ACS President Joseph S. Francisco's Presidential Task Force on Innovation that chemical entrepreneurs need more training and affordable resources to plan, launch and grow chemistry-based start-ups. The pilot EI was a good start, but several lessons were learned from pilot program participant feedback, namely that:

1. Information about entrepreneurship is good, but hands-on instruction for creating a viable, sustainable business plan is better;
2. Understanding the terminology of entrepreneurship is good, but knowing how it applies toward achieving an investment is better;
3. Knowing what to do is good, but doing it is better;
4. Networking with investors or partners is great, but due diligence and term sheet negotiations for funding events are better;
5. Knowing what your technology does is good, but knowing what the customer wants and will pay for is better;
6. Having colleagues on your management team is good, but having a knowledgeable, experienced, and capable executive team is better; and
7. Concepts are great, but working prototypes are best.

These and other lessons have inspired and guided ACS efforts to assist chemical entrepreneurs. Future refinement and retooling of the ACS Entrepreneurial Initiative will better address the challenges associated with the most pressing, near-term need of nearly all chemical entrepreneurs, namely funding. Improved opportunities for job creation by these chemical entrepreneurs are more likely to be achieved from investments into de-risked technologies and start-ups.

Acknowledgments

Many thanks to the members of the Entrepreneurial Initiative Task Force who provided advice, direction, and assistance during the implementation and evaluation of the Pilot – Entrepreneurial Initiative.

References

1. Voith, M.; McCoy, M.; Reisch, M. S.; Tullo, A.H.; Tremblay, J. Facts & Figures of the Chemical Industry. *Chem. Eng. News.* **2010**, *88*, 33–67. Pages 50-53 emphasize layoffs in the chemical and pharmaceutical industry between 1999 and 2009.
2. *National Occupational Employment and Wage Estimates United States, 1989-2009*; Technical Report; U.S. Department of Labor, Bureau of Labor Statistics: Washington, DC, 2009.

Chapter 14

Creativity, Innovation and the Entrepreneurship Nexus

Sadiq Shah*

The University of Texas Pan American, 1201 W. University Drive, Edinburg, Texas 78539-2999
*E-mail: sadiq@utpa.edu

Creativity and innovation are the driving forces behind new and unique ideas for improving existing process, identifying totally different solutions to problems, developing novel products or services, identifying and establishing new and novel platforms for ideas that were not conceived of before. Despite our apparent similarities based on our genetics, our individual experiences lead each of us to think differently. The diversity of ideas and solutions in any society are the collective product of its individuls' unique experiences. Some of these ideas can cross the nexus to the next step of entrepreneurship, which translates ideas into desirable products and services. Successful entrepreneurship generates financial rewards, and this fact offers incentives for creative minds to participate in this step of the process. This chapter will explore the role of creativity, innovation and the importance of entrepreneurship to translate ideas into desirable products for the society.

What Is Creativity?

Creativity is about finding a new and a different way of thinking and/or doing things that are beyond the existing practice, and no one else has considered. It is looking for that fresh approach that no one else is. The creative idea may be based on the synthesis of existing knowledge in a way that it creates something different or it may be based on totally different and a novel approach. The outcome of creative thinking includes examples such as dance, music, poetry

© 2014 American Chemical Society

and literature represent interpretations inspired by emotions, feelings, rhythms and pastures, or it may be inventions or technological innovations connecting existing knowledge or new knowledge to create products or services that are totally different than the current practice. It may involve asking questions in ways that expands our perspectives, and advances our knowledge base and facilitates identifying different solutions to existing or new problems. What entails the creative process? It is about allowing our imagination to think beyond any defined boundaries (outside the box) and not necessarily constrained by our imagination intentionally or unintentionally by the context of one's life experiences, the limits of educational background, and the boundaries of intellectual perspectives. The creativity is inspired by the environment, the stimuli to our imagination and the tendency to reframe the problem, the questions asked and the approach to solving it. It is the free thinking that inspires the imagination to conceive ideas and approaches that are otherwise unexpected, different, novel and impacts the human perceptions with a response of wow, this is unique.

Imagine if we were all thinking alike; the world would certainly be different (1, 2) than what we live in today, but certainly stagnant. Most likely we would not have made advances in knowledge base and perhaps the quality of life may not have improved much more than the primitive times. It is the fundamental creative genius of human beings that drives the generation of ideas, new and improved and breakthrough solutions to societal challenges. Each generation makes an advance in various fields through their creativity. The next generation builds on it or pursues a different direction through a breakthrough. It is well known that we cannot expect a different result by doing the same thing over and over.

Creativity dynamics illustrated in Figure 1 is aiming at bringing about a paradigm shift. By using creativity either by connecting discovery with a need in the society through serendipity, or using the creative approaches to link innovation to meet the need (3). This innovation may be the result of bridging discovery through foresight to innovation that connects with societal need through creativity.

The Process for Creativity

Incremental Ideas

Creativity can be based on small steps and existing knowledge. This type is often based on vertical thought process. Vertical thought process is based on existing ideas and builds on an existing framework, it is low risk with some history of success; however, it never results in breakthrough ideas or solutions. The analogy is drilling deeper into an existing well for more water when the well dries up, hence a less risky and less costly approach to potential access to more water.

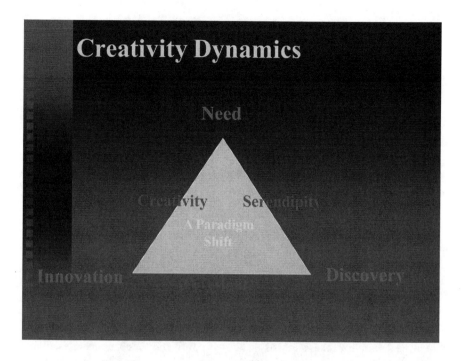

Figure 1. An illustration of the creativity dynamics

Breakthrough Ideas

Creativity can also be based on discontinuity or a total disruption in our usual pattern of the thought process, beyond the traditional way of looking at the problem, a totally different and an unusual approach to solving the problem. This type of creative process utilizes horizontal thinking; each idea is a totally different than the previous idea. It is neither a linear thinking nor sequential, each approach is unique and the creator is not satisfied with one idea and is continuously looking for a different and a novel approach. These ideas are based on divergent thinking, where ideas are not judged initially by necessarily the practicality but facilitated by the synthesis of ideas with a totally different frame of mind. The analogy here is drilling horizontally for a new wells to access more water, hence a much more risky yet a bold approach; however, and it is most likely to lead to a breakthrough ideas (*4*).

No one approach is right and the other wrong, it is what the situation demands or the human imagination takes us. Depending upon the problem or the situation under consideration, it is often times a hybrid approach that can be productive.

Creativity is not a linear process; it is based on discontinuity and often a quantum jump. The thought process is generally not a planned and a deliberate process it is spontaneous in response to something that stimulates our imagination. The creative thinking is often driven by reframing the problem, the issue that allows our imagination to look at it with a different perspective. Often the environment and external stimuli make a difference in inspiring our imagination which leads to creativity. It is about looking at the problem asking questions in different ways, the "why" and "what if" type of questions. The creative process may start with random thinking with diffused chaos, which is a formative stage. It is then followed by some sorting to recognize different patterns that may result in some focused ideas or solutions that ultimately translate into novel ideas that can be implemented.

Creative ideas can be the result of evolution of incremental changes for improvements on existing practices, as in vertical thinking. Alternatively, in horizontal thinking, it can be based on the synthesis of two or more totally different ideas through a new and different pattern of the thought processes, and this thought process may lead to a complete breakthrough; it may be a revolution of ideas by exploring totally different approaches and perspectives. It may also involve a reapplication of existing ideas beyond what is being practiced and thereby creating something new.

The creative process is driven by self motivation, the boundaries of our knowledge and the limits of our imagination to create a new idea often based on the building blocks of our existing knowledge. The synthesis of creative ideas is often triggered by oral and/or visual clues to stimulate our senses thereby generating new ideas. Think about the thought process that young children go through, it is uninhibited, and they turn their imagination loose and the free thinking, outside the box without any judgment. All human beings are creative; we just need to allow our imagination to be freed with no constrains for conformity to the norms. Creativity can be a spontaneous process triggered by something else; however, it can also be a deliberate process as it is often done through a brainstorming process. Where ideas are generated with no regard for the practicality, and often ideas are synthesized in the process to come up with the final solution. If human beings were not inherently creative our world would have been at stand still.

What Is Innovation?

Creativity is driven by our imagination. But imagination is not enough to make an impact for the society. Innovation is the step to implement and translate imagination and creativity into a meaningful outcome that either an individual benefits from intellectually or professionally or personally, and/or the society benefits from in a meaningful way. This can be a product or a service that fulfills a need in the society.

Innovation is the implementation of the outcome of creative process. Not every creative idea can be translated into a product, service a process or a new way of looking at the problem, and not every creative person has the ability to implement their ideas. The use of a creative idea is possible only through the creativity, innovation and the entrepreneurship nexus. Thus an incremental creative idea will lead to incremental innovation, continuous improvement of existing art. Whereas, a breakthrough idea will lead to a radical change, disruptive technology with discontinuous improvement in existing practice, hence a breakthrough innovation. Examples of disruptive innovations include "Post-it®" by 3M. In 1963 Spencer Silver discovers a low tack adhesive that leaves no residue, that is strong enough to hold the paper together but weak enough to pull apart. Art Fry during his church choir practice frustrated with the book mark falling repeated to the ground. During one sermon, Fry recalls his colleague's discovery of low tack adhesive and imagines the use of low tack adhesive for a bookmark and the rest is now history. The low tack adhesive became the vehicle for this innovation and through entrepreneurial initiative 3M launched a product that significantly impacted and enabled our daily routines in work places in particular.

Figure 2 shows the quantum jump from Walkman® audio tapes to Discman® to MP3® players to iPods® and finally to iPhones® or the Smart Phone and breakthrough innovations from the first generation of heavy hand held mobile phones to now the light iPhones® and Smart Phones are other examples of the disruptive innovation and entrepreneurship nexus. It is the creatively through cross-disciplinary thinking and connecting the discovery with the need, and/or the use of creativity to link serendipity through innovation to the need and the next step of entrepreneurship that facilitated the evolution of these unique products.

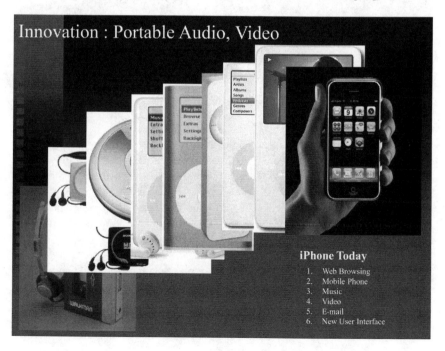

Figure 2. The evolution of technology for portable audio devices

Figure 3 below is an illustration based on a typical university's organizational structure. One can imagine potential innovations through collaborative interactions between various disciplines in different colleges with disciplines in a College of Science and Engineering and identify platforms for innovations for unique applications.

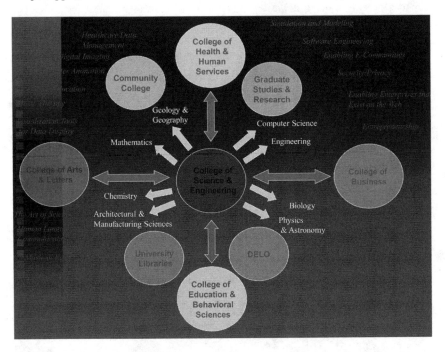

Figure 3. Pictorial representation of possibilities for cross-disciplinary technology development.

The goals behind creativity and innovation are to discover new frontiers, seize new opportunities, and apply creative thinking for innovative applications, asking questions that have not been asked to expand the horizons of human knowledge and thereby helping human endeavors and changing lives. Innovation can also be through the strategy for a process, operation, an approach or a new technology to impact the organizational effectiveness. Some practical examples of innovation include:

- An innovative product, service or marketing approach to create a novel and a differentiated product, service or market strategy
- An innovative business model that significantly changes the financial outcome of the business
- An innovative approach to business processes that dramatically changes the critical organizational performance

Apple Company's iPods®, iPhones® are examples of breakthrough technologies, and iTunes® for iPods® are examples of breakthrough marketing strategies. Organizations are always looking for ways to distinguish themselves in the market place whether they are for profit businesses or non-profits including educational institutions.

As illustrated in Figure 4 innovation may go through different stages such as the formative stage when ideas are chaotic, leading to well defined ideas, a normative stage when certain patterns can begin to emerge, followed by reaching a maturity stage where patterns are very well defined and the innovation has reached the point of diminishing returns as illustrated in Figure 5.

However, this may be the stage that can lead to a transformative change where new patterns emerge that offer the opportunity to take it through the next evolution of the innovation and potentially a breakthrough.

All human beings have a personal intellectual need to feel fulfilled in their ability to make a difference for the society, by making a contribution that brings personal gratification and intellectual growth.

Creativity is about weaving the threads of ideas through often a random process to create something totally different that most people may not think about. This creative process may play on the synergy of what we know and create something that we do not know. Creativity is not an inherent human ability, but self taught practice and the need and the desire to look for ways that are different. Creative people allow themselves to not settle on the first idea they conceive, they keep looking for ways that are different.

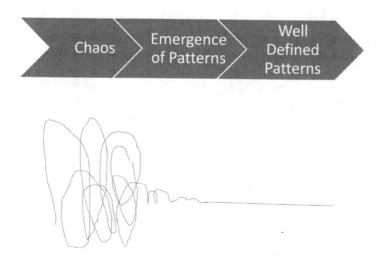

Figure 4. A representation of random thoughts evolving into well defined ideas.

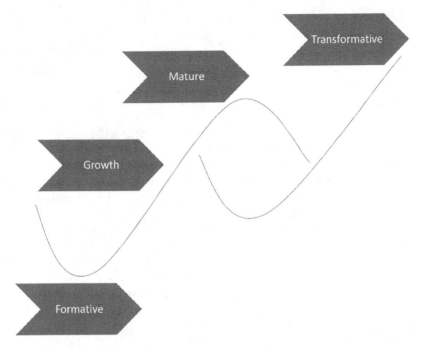

Figure 5. Growth of well defined ideas with time.

Creativity is a basic human intellectual need. Each human being is unique and through intellectual capacity, through physical identity and other attributes. We may or may not realize it, but we all have an inherent need to make our unique personality stand on its own ideas, different approaches, impressions, solutions, outlook, and perspectives to make our mark on the world, regardless of how big the definition of our world might be. For some, our world may be small, just our family and friends, for some it may be a larger circle including our acquaintances and colleagues and for the rest of us it may be the larger world.

To paraphrase Thomas Edison, the world outside of your industry, market or profession is full of existing ideas that people have never fully capitalized upon, which may be adapted to your specific need or challenge.

Entrepreneurship

Entrepreneurs are individuals who see opportunities in challenges while other do not. These entrepreneurs are the bloodline for the U.S. economy. The U.S. Small Business Administration defines a small business that has less than 500 employees. The majority of the businesses in the U.S. are small businesses.

According to the Small Business Administration 2012 publication related to the frequently asked questions, in 2010 there were 27.9 million small businesses (99.93%) while only 18,500 businesses (0.07%) had 500 or more employees (5, 6). According to the Bureau of Labor Statistics Business Dynamics Data (BED) from 1990 to September 2005 firms with less than 100 employees contributed an average job gains of 61.4 percent, while firms with less 500 employees generated 77.2 percent of the job gains.

A large number of the international firms as we know them today were born as small businesses by one or more individual entrepreneurs with an idea and a dream. More importantly with a plan and with their persistence they grew these firms to large corporations and successful global companies. Often times individual entrepreneurs are sitting on the fence while considering the potential launch of the business. The past recessions in a number cases played an important role to encourage these entrepreneurs to make their decisions about the launch of their businesses. During recession employment is down, labor, rent and supplies are cheep, such an environment catalyzes the entrepreneurial decisions for the birth of new small businesses. Figure 6 below illustrates examples of large corporations that we know today were born during the past recessions. In all cases creative minds with novel ideas and experience when faced with corporate downsizing took an entrepreneurial path to translate their ideas into products.

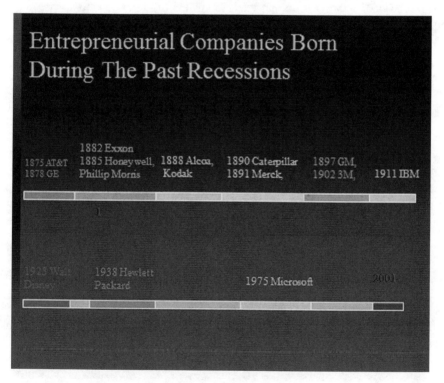

Figure 6. Corporations born during the past recessions.

The U.S. economy started out as agriculture based economy that evolved with the industrial revolution in the 1940s and on. In the 1970s information technology gave the economy the next jump followed by the Biotechnology in the 1990s. Each technology development served as a technology platform to support the applications and development of new products and services by entrepreneurs in small businesses and the scientists and engineers in large corporations. Each technology platform gave the economy a significant boost with investment and the creation of new jobs. It appears that the next frontier may be the Brain Research through Advancing Innovative Neurotechnologies (BRIAN) initiative that the current administration is giving a priority (7). This will take a sizable investment and it can engage scientists and engineers from a broad spectrum of disciplines.

So what states dominate in technology based jobs? Based on the data from the U.S. Patent and Trademark (USPTO) California appears to lead in the number of companies engaged in research, it ranks number one among the states that the USPTO grants patents to inventors who reside in California (8). The state that ranks number 2 is Texas followed by New York. This is also supported by the National Science Foundation Data that shows that California ranks number one in Science and Engineering employment base, Texas is number two and New York number three (9, 10). However, California offers a broad spectrum of technology based jobs because a broad spectrum of technology based companies have headquarters in California, while Texas the Science and Engineering jobs are predominantly in the Information Technology and Engineering markets with some presence in life sciences, while New York offers jobs in a range of disciplines including a significant number in life sciences.

R & D Performance

	Share of All R&D	Industry	Federal Government	Academia	Non-Profits	Total
Basic Research	18%	20%	7%	60%	13%	100%
Applied Research	22%	72%	8%	13%	7%	100%
Development	60%	91%	6%	1.5%	1.5%	100%
All R&D	100%	72%	8%	16%	4%	-

Figure 7. NSF data illustrating R&D performers in different roles. (Reproduced with permission, Copyright 2012. Battelle, R&D Magazine - Global R&D Funding Forecast)

It is helpful to look at the data generated by NSF based on the Research & Development performer. Figure 7 illustrates who the major players are in basic, applied and development efforts when it comes to the technology areas.

Since World War II, as a nation U.S. recognized that academia has a lot of talent that needs to be taped and it can also serve as a pipeline for producing future scientists and engineers to continue to maintain the U.S. leadership role in technology development and to remain competitive in the global market. Despite what we may hear in the media, U.S. higher education is still the envy of the world and a large number of international talented students come to the U.S. educational institution for advanced degrees. The U.S. government makes a sizable investment in research compared to the rest of the world. More importantly a significant portion of the funding from various federal agencies flows to the U.S. educational institutions for developing innovative curriculum and research activities to engaging graduate students and postdocs. Without this funding the number of students pursuing graduate education in STEM disciplines and the faculty and student based collaborative research and scholarship model will not function. According the report published by the Battelle Corporation, the data shows that the U.S. leads in the number of scientists and engineers per million people followed by Japan and China (*11*). U.S. also leads in the relative size of Research & Development annual expenditures. Figure 8 below illustrates the relative Research & Development expenditures over the last three years. The economic growth model that has emerged in the U.S. is based on the research sponsorship by the federal government, the early stage technology from academia is then transferred to the private sector and the private sector then leverages its core competency, the infrastructure and the financial resources to commercialize the technology. Over the last 60 years 50 percent of the economic growth in the U.S. has been the result of the federal investment in basic research and development efforts. It has served as a catalyst for the creation of new industries and jobs. The federal government will continue to invest in basic research in promising fields and create policies to facilitate the transfer of intellectual property to the private sector (*9*). Figure 9 shows the U.S. federal investment in the research and development over the last several decades based on the national priority at different times. In the 60s there was a heavy investment in space research (yellow), in the 70s in energy (green), in the 80s in information technology (red) and in the 90s in life sciences or biotechnology (blue).

The Association of University Technology Managers reported that in 2010 academia reported $59.1 billion total sponsored research expenditures, which includes $39.1 billion federally funded sponsored research expenditures and $4.3 billion in industry sponsored research expenditures (*12*).

The private sector, based on the market incentives, is well positioned to commercialize intellectual property into products and services. This model will continue to create jobs to further catalyze economic growth. Figure 10 below illustrates the Innovation cycle that has emerged in the U.S. so far.

Share of Total Global R&D Expenditures

	2010	2011	2012
Americas	37.8%	36.5%	36.0%
U.S.	32.8%	32.0%	31.1%
Asia	34.3%	35.5%	36.7%
Japan	11.8%	11.4%	11.2%
China	12.0%	13.1%	14.2%
India	2.6%	2.8%	2.9%
Europe	24.8%	24.5%	24.1%
Rest of the World	3.09%	3.1%	3.2%

Figure 8. Global Research and Development expenditures. (Reproduced with permission, Copyright 2012. Battelle, R&D Magazine – Global R&D Funding Forecast)

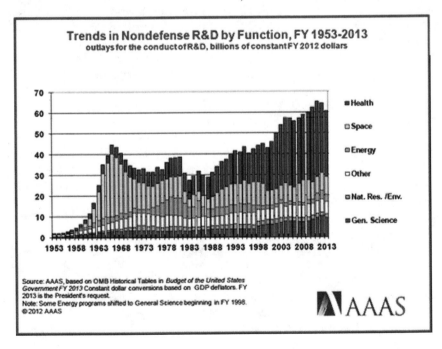

Figure 9. Trends for the U.S. federal research and development investment based on the national priority. (Reproduced with permission, Copyright 2013. American Association for the Advancement of Science)

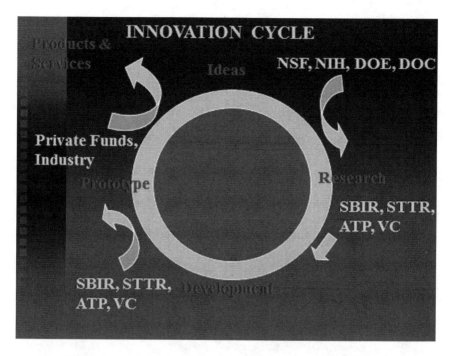

Figure 10. An illustration of the innovation cycle and the different stages of the development and funding sources.

Conclusions

Investment in research and development in science and engineering plays a critical role in bringing about innovation and through entrepreneurship catalyzing economic growth. However, innovation by itself is not sufficient to support economic growth. An essential driver for this economic growth is a pipeline of entrepreneurs. Thus innovation and entrepreneurship nexus based on the human creativity and persistence with appropriate knowledge of the intellectual property protection laws and the commercialization milestones will accelerate the U.S. economy over the next century. However, educational reforms to include the basics of the intellectual property laws and the business concepts for scientists and engineers will accelerate economic growth and also strengthen the U.S. technology based leadership role in the global competition.

Acknowledgments

Thanks to the American Chemical Society Presidential symposium organizers for creating a platform for discussion on this important topic and the ACS staff for their tireless support for making this dialogue possible. Special thanks to Dr. Robert Rich, Frank Walworth and Max Saffell for their hard work and support for the presidential symposia.

References

1. Csikszentmihalyi, M. *Creativity: Flow and the Psychology of Discovery and Invention*; HarperPerennial: New York, NY, 1997.
2. Florida, R. *The Rise of the Creative Class*; Basic Books: New York, NY, 2012.
3. Nuhfer, E. Embracing Cinderella: Beginning an Essential Conversation on the Metadiscipline of Technology Educating in Fractal Patterns XXXVI. *National Teaching and Learning Forum* **2012**, *21* (6), 77.
4. DeBono, E. *Lateral Thinking: Creativity Step by Step*; Harper & Row: New York, NY, 1970.
5. Helfand, J.; Sadeghi, A.; Talan, D. Employment dynamics: small and large firms over the business cycle. *Monthly Labor Reviews* **2007** (March), 40–50.
6. *SBA Frequently Asked Questions, Advocacy: The voice of small business in government*; 2012; pp 1–4.
7. Brain Research through Advancing Innovative Neurotechnology (BRAIN) Initiative, National Institute for Health. http://www.nih.gov/science/brain/ and http://www.whitehouse.gov/blog/2013/04/02/brain-initiative-challenges-researchers-unlock-mysteries-human-mind.
8. Patents by Country, State and Years – Utility Patents, December 2012. http://www.uspto.gov/web/offices/ac/ido/oeip/taf/cst_utl.htm.
9. *InfoBrief*; National Center for Science and Engineering Statistics, NSF, August 2013.
10. NSF data on Regional Concentration of Scientists & Engineers in the United States. http://www.nsf.gov/statistics/infbrief/nsf13330/.
11. 2012 Global R&D Funding Forecast. *Battelle R&D Magazine*; December 2011, pp 1–35.
12. AUTM 2010 Annual Survey report highlights.http://www.autm.net/AM/Template.cfm?Section=FY_2010_Licensing_Survey&Template=/CM/ContentDisplay.cfm&ContentID=6874.

Chapter 15

Innovation and Entrepreneurship in the Chemical Enterprise

Pat N. Confalone[*]

Director, American Chemical Society, Confalone Consulting, LLC,
303 Centennial Circle, Wilmington, Delaware 19807-2131
*E-mail: confalone@comcast.net

The important distinction between invention and innovation is presented. The ability to innovate and create start ups, even in the face of challenging economic environments, is illustrated from an historical perspective. Translation of real innovation into job creation and the critical role of the entrepreneur in this process is discussed. The wide range of opportunities in the twenty-first century for innovation in chemistry as the enabling science addressing major global challenges is presented. The scope of the ACS Entrepreneurship Initiative and Training programs, which afford critical assistance to budding entrepreneurs, is discussed.

Two former CEOs, Bill Gates of Microsoft and Chad Holliday of DuPont, wrote in an April 23, 2010 OpEd in the Washington Post: "The core force of innovation – vision, experimentation, and wise investments – has led to thousands of breakthroughs that benefit us all. A serious commitment to innovation can be transformative." Indeed, the well-established linkage of R&D investment and its translation to societal benefit, a vibrant economy, an improved quality of life, and job creation has been demonstrated for centuries. This chapter will focus on the two critical intervening steps between investment in R&D and the above outcomes, namely invention and innovation.

Wikipedia defines invention as "a new composition, device, or process derived from a pre-existing idea, or independently conceived in which case it may be a radical breakthrough". Albert Szent-Gyorgyi famously stated, "Discovery consists of seeing what everybody has seen and thinking what nobody has

© 2014 American Chemical Society

thought". However, invention although necessary is not sufficient for translation to the marketplace and the ensuing societal benefits. "Build a better mousetrap and the world will beat a path to your door", as Ralph Waldo Emerson has been quoted, misses the key next step in the process that must follow invention – the requirement for innovation. Referring once again to Wikipedia, innovation is defined as "an idea successfully applied to bring a value proposition to the marketplace and society. It is distinguished from invention, which are ideas made manifest". In other words invention is the conversion of cash into ideas, whereas innovation is the conversion of ideas into cash. Consider Nikola Tesla. He was an inventor and invested a lot to create his inventions, but was unable to ever monetize them. In contrast, Thomas Edison, although the holder of over one thousand patents, also drove innovation to the marketplace, amassing a fortune while bringing incredible products to the public. We all learned at an early age that Elias Howe invented the sewing machine. Today we continue to employ Singer rather than Howe sewing machines because it was Isaac Singer who commercialized this invention and built a hugely successful business. Invention may occur in the laboratory, but innovation must take place in the market. Innovation is a highly creative process that must include all business functions to be successful. An historic list of notable innovators would include Charles Goodyear [1839, vulcanization of rubber], Louis Pasteur [1862, pasteurization], George Eastman [1885, film], Charles Hall [1888, aluminum], William Hershey [1900, milk chocolate bar], William Burton [1912, catalytic cracking], and George Washington Carver [1925, peanut products].

In Disney World's EPCOT, "The American Adventure" features a large statue of a chemist holding up a test tube gazing intensely into its contents in front of his lab bench. The work is entitled "Spirit of Innovation" (Figure 1).

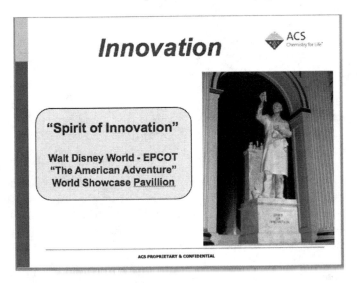

Figure 1. A statue of a chemist holding a test tube outside EPCOT. Courtesy of the American Chemical Society.

Table 1. Selected Chemical Innovations by Decade

1930-1940	poly(vinyl chloride), polyethylene, synthetic rubber, nylon, methyl methacrylate, poly(vinyl butyral), cellulose acetate, ion-exchange resins, invisible transparent glass, glass fibers, Vinyon
1940-1950	styrene-butadiene, poly(vinylidene chloride) (Saran), acrylic fibers, synthetic detergents, color photographic film, insecticides and herbicides, fluorocarbon refrigerants and propellants, tetrafluoroethylene polymers (Teflon), styrene-divinylbenzene resins, metal-coated fibers, Modacrylic, titanium dioxide
1950-1960	acrylic fibers, polyesters, polyester films and fibers, adhesives, ethylene oxide production of acrylics, low- and high-density polyethylene, ABS polymers, Spandex, olefin fibers, tetraethyl lead
1960-1970	propylene-ammonia production of acrylics, Surlyns, Hypalon, Lexan, Delrin, Celcon, PPO, paralenes, polysulfones, Anidex fibers, superconductors, Kevlar
1970-1980	conductive polymers, dendrimers, statins, PCR techniques
1980-1990	fullerenes, nanomaterials, HIV protease inhibitors
1990-2000	industrial applications of ionic liquids, metallocene based polyethylene
2000-2010	biotech synthesis of 1,3-propanediol)PDO) from corn, thin films of cadmium telluride or copper indium gallium selenide

We shall see that the opportunities for innovation in the chemical enterprise have never been better! Innovation must be market driven, guided by accurate insight and foresight. The perceived opportunity has to be data driven and fact based. The innovation process must be stage gated as it progresses to the marketplace and, in today's environment, sustainability focused. It is not sufficient to "do the work right" which means excellent science and collaboration, etc. but on must also "do the right work", hence the important of market driven innovation. Strive for relevance and uniqueness in the unmet market need that is targeted, ensuring that all aspects of the journey from "eureka" to the customer as resourced to win. True innovation begins with market need, succeeds through collaboration, and really doesn't count until you get paid. In summary, identify opportunities based on deep market knowledge, select the best opportunities for the business, integrate with core and/or accessible competencies, and manage and execute the projects with a clear intent to win. These series of state gated steps constitute a deliberate innovation process and will successfully drive invention to the market.

Too often unmet market needs continue to remain unmet because of a resistance to change, thereby creating opportunity for the entrepreneur. Henry Ford in his 1922 book *My Life and Times* said, "Businessmen go down with their businesses because they like the old way so well, they cannot bring themselves to change."

An examination of chemical innovations by the decade reveals a disturbing downtrend in the number of disruptive breakthroughs (Table 1).

This downtrend was a key finding of the ACS Presidential Task Force report *Innovation, Chemistry, and Jobs* (Figure 2), which can be found at www.acs.org/CreatingJobs. A second key finding was that start ups and small to medium sized companies create about 3 million jobs a year, in contrast to the 2 million that are lost in large companies. This surprising fact held true in all sectors of the economy over a span of several decades (Figure 3).

Figure 2. Front cover of the ACS Presidential Task Force Report on "Innovation, Chemistry, and Jobs". Courtesy of the American Chemical Society.

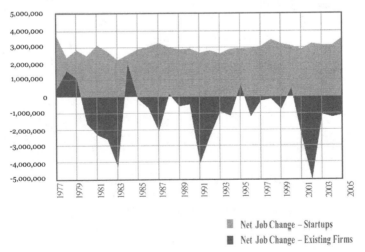

Figure 3. Job creation vs. job loss, 1977-2005. Courtesy of the American Chemical Society.

It's clear that the role of the entrepreneur in job creation is not only essential but also has been key to our robust economic growth. Of course, all the fortune 500 companies were at one time start ups, e.g. DuPont, founded in 1803, which basically manufactured gunpowder and did little else for the nineteenth century. In the middle of the deep recession of 1973-4, the following companies were start-ups: FedEx, Microsoft, Oracle, and Southwest Airlines. Bill Hewlett and David Packard started their company in a Palo Alto garage in 1939 investing the grand sum of $ 538. In the midst of the great panic of 1837, William Proctor, a candle maker, and James Gamble, a soap-making apprentice, joined forces. P&G had sales revenues of $ 82.6 billion in 2011. So even in tough economic times, the entrepreneur has risen to the occasion and created great companies that have prospered long after their early struggles.

A third key finding of the task force is that about 300,000 U.S. based jobs in chemistry have been lost over the last twenty years, in sectors that are unlikely to recover (Figure 4).

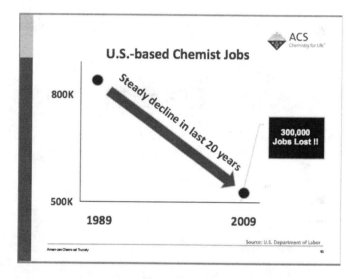

Figure 4. Trends in the U.S.-based chemistry jobs, 1989-2009. Courtesy of the American Chemical Society.

So, how do we win in these uncertain times? Consider the following facts which are the basis for maintaining an optimistic view of our future: 1) The United States has the best research universities in the world. This global leadership is unlikely to be seriously challenged in the next decade. 2) There are vast sums of angel and venture capital available. California, for example, has more venture capital than any country [except, of course, the U.S.] 3) We have a long tradition of science and technology driven innovation. 4) Centuries of U.S. entrepreneurship have led to the great companies of today, as a non-exhaustive set of examples shown above demonstrates. Moreover, there is now an unprecedented need for chemical innovation. Most of the start ups in the past several decades have been in biotechnology, driven by the continuing requirement for new medicines for the treatment of a myriad of unmet medical needs. These drivers will continue, of course, but consider the major challenges facing the planet as we move forward in the 21st century (Figure 5).

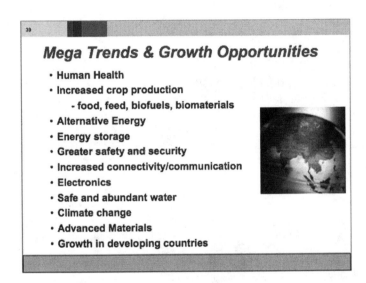

Figure 5. Mega trends and growth opportunities for the future global chemistry enterprise. Courtesy of the American Chemical Society.

Each one of the global challenges in this list of megatrends will require chemical innovation as a key component of the potential solutions. Opportunities to impact human health through new medicines and medical devices, increasing crop yields, developing alternative energy, improved energy storage and battery technologies, communication, electronics, the potable water challenge, climate change, advanced materials – all demand the application of chemistry across this palate of global opportunities. Even today, 96% of all industrial processes have a chemical component, requiring the need for improved processes as well as products – all solved by disruptive or sustaining innovation throughout the chemical enterprise.

The final key finding of the task force reinforces this view of great opportunities for entrepreneurship. Chemistry related R&D expenditures in companies are shifting from large companies to businesses with under 1,000 employees, a fact that spans three decades (Figure 6). Even as big pharma, which for over fifty years has been among the most innovative and profitable branches of the chemical enterprise, continues to undergo unprecedented contraction, biotech start-ups and development stage companies continue to innovate and create jobs.

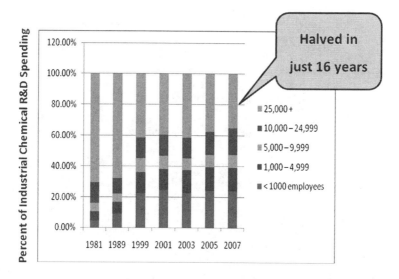

Figure 6. Industrial R&D spending by U.S. chemical companies by employee number, 1981-2007. Courtesy of the American Chemical Society.

In this environment of the new realities, called by some the new normal, in which expectations, threats, and opportunities have been reset, four conclusions can be made:

1) There is no intrinsic structural reason to believe that chemistry start-ups could not prosper in chemistry under the right circumstances, even in difficult economic times
2) There is an opportunity for the chemical enterprise to rebuild its leadership position through disruptive technology, enabled by innovative start-ups
3) Early stage companies provide a promising opportunity to create new jobs for innovative young chemists and seasoned professionals alike
4) There are major opportunities for the ACS to facilitate entrepreneurship and enable the creation of new companies, thereby creating high quality careers for U.S. chemists

Clearly the ACS has a role and opportunity, even an obligation to enable our members to initiate and execute entrepreneurial activities. For most of our history, the focus has been on the large chemical companies that comprised the bulk of our industry-based membership. It's now time to turn the dial toward the creation of new chemical businesses and assist our members in this timely and important endeavor. This can be accomplished in four major thrusts:

1) Help entrepreneurs create jobs by facilitating more affordable access to key resources like information, expertise, important services for start ups [HR, finance, etc.], and mentorship

2) Improve the business environment for startups by increasing advocacy of relevant policies at the federal and state levels
3) Partner with academic institutions to promote awareness of career options that involve entrepreneurship
4) Publicize the challenges and successes of entrepreneurs and entrepreneurship in the chemical enterprise

Indeed, implementation of all of these recommendations has begun. In particular, the ACS Entrepreneurship Initiative, moving beyond its pilot stage, is an important advance and consists of both a training component as well as a resource center for budding entrepreneurs. These programs are designed to help overcome the barriers that exist en route to the all important seed funding that launches a start up company. Hurdles consist of understanding the regulatory environment, intellectual property concepts [and expenses incurred in filling and maintaining patents], building a credible business plan, market insight, etc. (Figure 7).

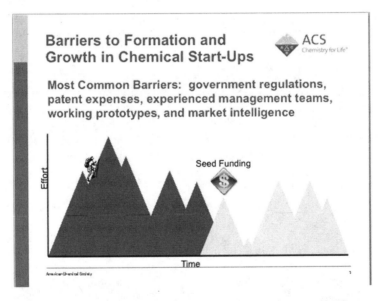

Figure 7. Barriers to formation and growth in chemical start-ups. Courtesy of the American Chemical Society.

The training component is called the ACS Entrepreneurial Training Program [ETP] and consists of course content licensed from the prestigious Kauffman FastTrac™ content. This experience provides business basics with real examples from actual ACS entrepreneurs and business professionals specializing in chemistry related start-ups, employing online support materials, including helpful video content. The second program, the ACS Entrepreneurial Resources Center [ERC] provides free or discounted access to important service providers, access to CAS SciFinder™ as well as ACS publications, and advice and mentoring from volunteer ACS entrepreneurs. In addition, the ACS has conducted both an Entrepreneurial Summit, which featured an educational forum, and Entrepreneurial Showcase events in La Jolla and Boston.

As we've seen, the opportunities for chemical innovation have never been better and the ACS is ready, willing, and able to assist our members to be a significant participant in bringing important inventions forward to their ultimate realization as true innovation. Society will reap the benefits of innovation enabled by entrepreneurship as the ACS continues to drive forward its mission to "*Improve people's lives through the transforming power of chemistry*".

Chapter 16

Venture Capitalist Planning Is Irrelevant to Successful Entrepreneurship in Chemicals, Materials & Cleantech

J. M. Ornstein[*,1] and Hongcai "Joe" Zhou[2]

[1]President, framergy, Inc. 800 Raymond Stotzer Parkway, Suite 2011A, College Station, Texas 77845, United States
[2]Professor, Department of Chemistry, Texas A&M University, 800 Raymond Stotzer Parkway, Suite 2011A, College Station, Texas 77845, United States
*E-mail: findoutmore@framergy.com

In 2010, Dr. Hongcai 'Joe' Zhou and cleantech commercialization firm J.M.Ornstein explored the possibility of licensing Metal Organic Framework materials funded by ARPA-E through Texas A&M University as the foundation for a new startup. The founders felt that the underlying technology was ripe for early commercialization and in support of US Federal funding goals, thought that the time was right to stimulate future jobs growth through entrepreneurism. But there are specific qualities of the chemicals, materials and cleantech (CMC) sectors that limit their appeal to venture capital by pressuring their J-Curve; these include high upfront research and development costs and the long-term nature of time to a liquidity event. To counter these challenges, the founders established a low cash burn plan for the framergy™ by leveraging university assets and making costs a priority. framergy™ was able to execute a market valuable license with Texas A&M University and strategic plan leading to seed funding of $250,000 in 2011 and Series A funding of approximately $1,600,000 in 2013. Many great frontier chemical and material sciences can reach the market through entrepreneurism if they match market realities with cost management.

© 2014 American Chemical Society

The J Curve

The J-Curve in sociology traces its origins to three economists from the early ninetieth century who independently sought to explain why devaluation of currency would first lead to current account deficit, and then would quickly swing positive. Alfred Marshall (*1*), Abba Lerner (*2*) and Joan Robinson (*3*) defined this condition where the sum of elasticities for imports and exports with respect to the exchange rate is greater than one.

$$(\varepsilon + \varepsilon^* > 1)$$

This effect, later known as the J-Curve based on its shape as expressed in figure 1, was adapted to country status by Ian Bremer (*4*), and revolution modeling by James C. Davies (*5*).

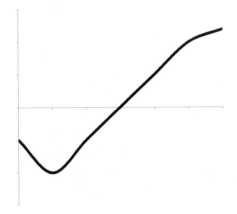

Figure 1. The J-Curve. (Courtesy of author).

In the last ten years, many economists and financial journalists began to see that this curve reflected a scenario for private capital investment where the sum of elasticities for investment and return with respect to capital employed is greater than one. Of particular note were *Exposed to the J Curve* by Ulrich Grabenwarter and Tom Weidig (*6*), and *Beyond the J-Curve* by Thomas Meyer and Pierre-Yeves Mathonet (*7*). By the late 2000's, the J-Curve became an essential tool to explain venture capital returns.

In venture capital (and private equity), the J-Curve X-axis represents time and the Y-axis represents the net of negative capital outflows and positive capital inflows. As the venture capitalist makes investments into one or more startups, the curve will be negative. This is not uncommon with startups regardless of sectors, but the relationship between loss and gain will be more pronounced in new technology. In consideration of a one firm investment, the net of negative capital outflows will be offset by earnings, with the point of inflection representing the maximum financial needs of the investment. This is when the firm is cash flow positive and the burn rate, the rate observed over a specified time, becomes less relevant to the venture capitalist. These venture capital attributes are associated to the J-Curve in figure 2.

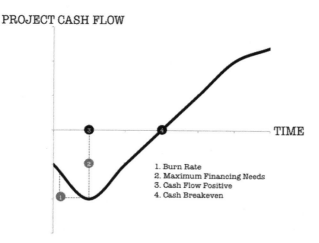

Figure 2. The Private Investment J-Curve. (Courtesy of author).

Venture capital funds typically have a term of ten years, with all returns, liquid or ill-liquid, returned at the end of this legal period for final accounting (8). Typically, within the first four years all funds committed by limited partners are drawn, with huge penalties imposed if the limited partner reneges (8). This puts pressure on the venture capitalist to invest the money, otherwise committed capital sits with low return and time exacerbates poor performance. Usually by year five, the venture capitalist would want to have all funds committed so that there will be at least five years to grow investments. Little has been innovated into this structure and each investments J-Curve reality will dictate a funds overall return.

With this understanding, the J-Curve reveals that a successful venture capitalist will seek startups where: 1) The date of cash flow break even is within five or six years, 2) the burn rate is low, 3) the maximum financing needs are lower when compared to like investments, and 4) where negative capital outflows are offset by aggressive earnings. Unfortunately, different industries have diverse financing needs, and different earnings potential over like time horizons. To encourage venture investments and/or limited partners, some firms may mislead the design of their J-Curve.

VC Market Realities

Since 2012 when the groundbreaking *WE HAVE MET THE ENEMY ... AND HE IS US* was published by the Erwin Marion Kauffman Foundation, an investor in venture capital funds, the realities of lower than expected returns from the venture capital industry were exposed (9). This report raised doubt as to why the industry was generally considered to offer outsized returns in the media. Clearly huge successes in the Dotcom bubble associated with Netscape, Google and EBay, and in the Web 2.0 period associated with Facebook, Groupon and Twitter, impacted media and encouraged the high risk, high return nature of early technology investing. "The structure of the venture capital industry, in which a small number of firms each needed to invest big sums in a handful of start-ups to have a chance at significant returns" (*10*) plays on the emotional nature of gambling versus the market realities.

It is true that venture capitalists since the 1960s appear to have had excessive returns (*8*), but this led to a crowding of the field. Unfortunately, over the last measured twenty years, 62% of venture capital funds failed to return to limited partners, their investors, returns higher than public markets (*9*). Yet entrepreneurs look to these firms as the holy grail of how to make a startup a financial success. Entrepreneurs should look at ten-year market returns (*11*) and the measure this against risk to understand a venture capitalist's financial risk profile. According to Cambridge Associates U.S. Venture Capital Early Sage Fund Index, end-to-end pooled return/net to limited partners for ten years is 7.36 percent (*12*).

To understand why the market realities of venture capital firms are poor, one must ignore the media hype and look at the statistics. One must also understand their structure and profit motivations to manage their 'advice' value proposition. For example, less than one percent of US companies raise money from dedicated venture capital firms (*13*). In addition, as new venture capital firms show success, they raise larger funds; this scaling is correlated to lower returns (*13*). Many venture capital firms become dormant, with some estimates of up to 400 dormant firms in the US (*14*).

The average venture capitalist is perceived as a risk-taker, on par with the entrepreneur, but who's deep experience and training will lead to advice on how to succeed.

> VCs (venture capitalists) are often portrayed as risk takers who back bold new ideas. True, they take a lot of risk with their *investors'* capital—but very little with their own. In most VC funds the partners' own money accounts for just 1% of the total. The industry's revenue model, long investment cycle, and lack of visible performance data make VCs less accountable for their performance than most other professional investors. If a VC firm invests in your start-up, it will be rooting for you to succeed. But it will probably do just fine financially even if you fail (*13*).

While much of the venture capitalist's personal return is associated with fees charged to their limited partners for advice and mentoring startups, the advice is driven by their J-Curve and not long-term results. This advice is often pitched to entrepreneurs during term sheet negotiation to favor the venture capital firm's returns over the entrepreneur. For the chemistry, materials and cleantech (CMC) entrepreneur, doing due diligence on offered advice is important.

Market Realities

Returning to the limited time that the venture capitalist has to demonstrate return, one can identify that there is a motivation to inflate revenue expectation. This motivation can come from the entrepreneur, venture capitalist, or become an unintended consequence of their cooperation. Ten years from startup to a liquidity event is very short, and most investments will not have the full ten years of the fund's vintage. In addition, the venture capital fund may return the investment to the limited partner in the form of securities if there is no liquid market.

Sales frameworks taught in business classes are based on strategies employed by mature firms. These firms don't have to develop product credibility and overcome customer-switching costs. This leads to an inevitable circumstance where the entrepreneur is lowering price as an incentive, eroding sales forecasts from year's prior. This is supported by a legitimacy curve (15), which pushes the J-Curve to the right on the X-axis, when the technology has a higher degree of adoption risk.

In venture capital, the legitimacy curve can be overcome through innovation leading to rapid desirability and substitution. This process was first described by Bryce Ryan and Neal Gross in 1943 (16), and then more precisely by Everett Rogers in *Diffusion of Innovations* in 1962 (17). Rogers went as far as to numerically pinpoint market penetration over successive sets of consumers. What allowed venture capital to integrate this into their J-Curve revenue expectation horizons was the impact of network computing on adaptation models. Network computing decreased the time to expose innovation to markets, but more importantly, allowed for transparent "contagion through direct network ties" (18).

By the late 1990s, the Dotcom sector was self-fulfilling this contagion and many thought all technology sectors could shift the J-Curve to the left. But the reality is that a decade later, the clusters of internet, software, mobile/telecom and electronics were the capital concentration of venture capital (19). The reality is that social networks could only overcome 1960s technology diffusion rates where perceived 'trialiability' and complexity (17) were not obstacles; which they are in CMC.

Market realities, irrespective of the final impact on the J-Curve, must be used in forecasting revenues. Every MBA knows to start revenue projections with the total size of the market. But the entrepreneur must measure reach, adoption and potential market share into revenue forecasting, as graphically depicted in figure 3.

Many startups don't measure new technology adoption, rather integrate it into market share. Rogers' 1960s era technology diffusion curve and other rational carve-outs will give the CMC entrepreneur a real tool to evaluate the worth of their investment opportunity.

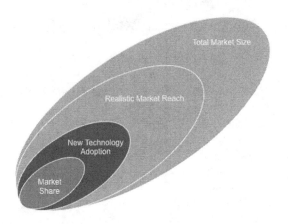

Figure 3. Ration Market Sizing. (Courtesy of author)

Cost Management

Again under the consideration of the limited time that the venture capitalist has to demonstrate return, another motivation would be to deflate financing need expectations through unreasonable cost projections. This motivation can come from the entrepreneur, venture capitalist, or become an unintended consequence of their cooperation. And again, the ten-year vintage of a venture capital fund will pressure this motivation.

The lean startup movement is gaining popularity; with failing fast as the ultimate solution to cash burn (*20*). Here, entrepreneurs are encouraged by venture capitalists to move closer to the consumer with the 'minimal viable product', thereby reducing the high cost of planning. Many researchers are being pushed in this direction by governmental grant requirements for tie-up with industry or support letters. In 2012, the National Science Foundation called upon lean startup proponent Steve Blank to design an entrepreneurship curriculum for scientists (*21*).

This lean startup process supports the venture capitalist's desire to shift the J-Curve up and to the left. But the entrepreneur in CMC needs to be cautious of the designs under the lean startup movement as they are intended to counter issues that challenge the J-Curve for internet and software startups. For example, 'agile development' encouraged in the lean startup movement is derived from software development (*10*), where product life cycle is far shorter than CMC.

Social networks and open source software can dramatically reduce research and development costs, but this type of resource is more readily shared in internet and software sectors. For CMCs, these resources are found in non-IP driven research, non-profit institutions and science incubators that understand the sector. But unlike software resources sourced on the web, this requires the CMC entrepreneur to be more proactive in engaging actual gatekeepers of resources. Often these gatekeepers are competitors with vast scientific and legal resources.

One good example of a lean start up would be virtual biotechnology companies, which surfaced in the 1990's to counter the costs of bringing new medicines to market. Many technology diffusion issues are shared between CMC and 'virtual pharmas': "These companies consist of as few as one full-time employee who oversees a drug from preclinical development to tests in patients, all in the hands of outside contractors" (22).

The CMC entrepreneur should also be cautious of venture capital micromanagement, which can increase total financing needs:

> As (Fred) Destin (of Atlas Ventures) explained, "Entrepreneurs often complain that VCs show up at meetings late, they micromanage, they are too busy to understand the subtleties of their business and they apply over-simplified solutions to problems" (23).

In fact, risk actually is measured to rise when venture capital firms apply more control (measured through board control or liquidity preference) (24). Virtual management and other rational cost management techniques such as efficient management will give the CMC entrepreneur a real tool to evaluate the worth of their investment opportunity.

Conclusions

Venture capital funds are known to "bury their dead very quietly" (25), but it has been revealed that their results differ from the media perception of extraordinary investment success. A small and contracting venture capital market (13) should not discourage the entrepreneur from forming a startup from new science in the CMC sectors. The venture capital motivations of the J-Curve may not exist for high net worth individuals and strategic corporate investors. For framergy™, this became clear at the annual 2012 ARPA-E summit, curated by the source of the governmental funding that financed much of framergy's technology. "Many VC (venture capital) firms parted ways with their cleantech teams in 2012. While February's (2012) ARPA-E conference had a record number of attendees, venture investors were scarce – replaced by a bumper crop of corporate types" (26).

The moral hazard of venture capitalists is that the financial risk of their investment is often shifted to their limited partners and entrepreneurs through deal structuring. framergy™ successfully closed a seed round and series A round to fund its CMC startup by ignoring motivations to manipulate its J-Curve to venture capital standards. Instead of seeking out venture capitalists, its entrepreneurs marketed the investment opportunity to the other 99% of US corporate capital providers. Frontier chemical and material scientists, and entrepreneurs, should avoid the perverse incentive of adjusting market realities and cost management to meet the needs of a select few.

References

1. Marshall, A. *Money, Credit and Commerce*; Macmillan and Co.: London, 1923.

2. Lerner, A. *The Economics of Control*; Macmillan: New York, 1944.
3. Robinson, J. The Foreign Exchanges. In *Readings in the Theory of International Trade*; Ellis, H., Metzler, L. A., Eds.; R.D. Irwin, Inc.: Homewood, IL, 1937; pp 83−103.
4. Bremmer, I. *The J curve: a new way to understand why nations rise and fall*; Simon & Schuster: New York, 2006.
5. Davies, J. C. Toward a Theory of Revolution. *Am. Soc. Rev.* **1962**, *27*, 5–19.
6. Grabenwarter, U.; Weidig, T. *Exposed to the J Curve: Understanding and Managing Private Equity Fund Investments;* Euromoney Institutional Investor: London, 2005.
7. Meyer, T.; Mathonet, P.-Y. *Beyond the J-Curve: Managing a Portfolio of Venture Capital and Private Equity Funds*; Wiley: Hoboken, NJ, 2005.
8. Shalman, W. A. The structure and governance of venture-capital organizations. *J. Financ. Econ.* **1990**, *27*, 473–521.
9. Mulcahy, D.; Weeks, B.; Bradley, H. S. *We Have Met the Enemy ... and He Is Us*; Erwin Marion Kauffman Foundation: Kansas City, MO, 2012.
10. Blank, S. Why the Lean Start-Up Changes Everything. *Harvard Bus. Rev.* **2013** (May), 64–73.
11. Cohan, P. Five Ways To Revive Venture Capital. *Forbes* [Online], June 6, 2013. http://www.forbes.com/sites/petercohan/2013/06/06/five-ways-to-revive-venture-capital/ (accessed September 2, 2013).
12. *U.S. Venture Capital Index and Selected Benchmarks*; Cambridge Associates LLC: Arlington, VA, 2013. (Full table is available online at http://www.cambridgeassociates.com/pdf/Venture%20Capital%20Index.pdf).
13. Mulcahy, D. Six Myths About Venture Capitalists. *Harvard Bus. Rev.* **2013** (May), 80–83.
14. Hogg, S. Beware of the living dead. *Entrepreneur* **2013** (September), 75.
15. Epstein, E. M.; Votaw, D. *Rationality, Legitimacy, Responsibility: Search for New Directions in Business and Society*; Goodyear Publishing Company Inc.: Santa Monica, CA, 1975.
16. Ryan, B.; Gross, N. C. The diffusion of hybrid seed in two Iowa comminutes. *Rural Sociol.* **1943**, *8*, 15–24.
17. Rogers, E. M., Ed. *Diffusion of innovations*; Free Press: New York, 2003.
18. Valente, T. W. Social network thresholds in the diffusion of innovations. *Social Networks* **1996**, *18*, 69–89, 85.
19. Steiner, C. Top 10 Sectors Venture Capital Likes Right Now. *Forbes* [Online], April 1, 2010. http://www.forbes.com/2010/04/01/venture-capital-trends-entrepreneurs-technology-venture-capital.html (accessed August 28, 2013).
20. Ries, E. *The Lean Startup: How Today's Entrepreneurs Use Continuous Innovation to Create Radically Successful Businesses*; Crown Publishing: New York, 2011.
21. Colao, J. J. Steve Blank Introduces Scientists To A New Variable: Customers. *Forbes* [Online], August 1, 2012. http://www.forbes.com/sites/jjcolao/2012/08/01/steve-blank-introduces-scientists-to-a-new-variable-customers/ (accessed August 25, 2013).
22. Ledford, H. Biotechnology: Virtual Reality. *Nature* **2013**, *498*, 127–129.

23. Cohan, P. Five ways to revive Venture Capital. *Forbes* [Online], June 6, 2013. http://www.forbes.com/sites/petercohan/2013/06/06/five-ways-to-revive-venture-capital/ (accessed August 28, 2013).
24. Kaplan, S. N.; Stromberg, P. Characteristics, Contracts, and Actions: Evidence from Venture Capitalist Analyses. *J. Financ.* **2004**, *5*, 2173–2206.
25. Gage, D. The venture Capital Secret: 3 Out of 4 Start-Ups Fail. *Wall Street Journal* [Online], September 20, 2012. http://online.wsj.com/news/articles/SB10000872396390443720204578004980476429190 (accessed August 25, 2013).
26. Nordan, M.. *Secret Formula* [Online Blog]. The State of Cleantech Venture Capital: What Lies Ahead. http://mnordan.com/2013/03/27/the-state-of-cleantech-venture-capital-what-lies-ahead/ (accessed August 25, 2013).

Chapter 17

A Tale of Academic Innovation to Job Creation in Rare Earth Extraction and Separation

Neil J. Lawrence,* Joseph Brewer, and Allen Kruse

Co-founders, Rare Earth Salt, LLC, 1111 18th Street SW Minot,
North Dakota 58701
*E-mail: chem@rareearthsalts.com

> In recent years it has become increasingly apparent that the global demand for rare earth elements is rapidly outstripping supply. Additionally it has recently been reported that there are few non-academic jobs for physical scientists with advanced degrees. This presentation will discuss how one group of colleagues saw these two problems and came up with a solution to both problems. Beginning with "out of pocket" funding for research in a garage, through initial fund raising, and selection of the right global partner for the construction of a pilot plant were each steps fraught with potential and real hurdles to be overcome. Each of the co-founders, and their families, had to be "all in" for this project to succeed. For everyone involved, there had to be a motivation greater than a simple paycheck; in order to start a new chemical enterprise you must consider the impact you will have on the future.

Introduction

Chemists are, almost uniquely, situated to have a wide variety of possible career paths. Because of their training in the understanding and manipulation of matter at the atomic level, chemists have the option of traditional career paths such as academia, or being a valuable employee in industry; or they can choose to strike out as entrepreneurs to start our own companies. In view of the rapid evolution of industry and modern technology and the explosion of global population in the 21st century, there are many emerging business opportunities to satisfy new global demands for chemicals products, from better laundry agents,

© 2014 American Chemical Society

to novel pharmaceuticals to cure new diseases, to advanced electronics materials in high tech industries. However, starting any new business from scratch in economically difficult times is considered hard, but starting a chemical business during a recession in 2010 was something that most would never consider doing.

In this chapter, I will share with you my experience starting my chemical business in 2010 after I earned a master degree in Chemistry at the University of Nebraska-Lincoln. After the completion of my master's degree in chemistry, I requested a pause in my academic studies to assist a friend in starting a business, during that time our training led us to realize that we had an opportunity to revolutionize rare earth ore refining; thus began our adventure. Having great mental preparation, endurance, and most importantly support from family are the critical elements for one to succeed in the initial stage of a startup business. In his book "*Aspire : discovering your purpose through the power of words*", Kevin Hall, defines the word "Ollin" as "to move and act now with all your heart" (*1*). This word is pronounced similarly to the English "all in" which describes the attitude that I believe everyone involved in starting a successful business must have. I recall a recent Christmas when I had to be away from my family for several months. During this period of time, my business partners and I were living in Minot, North Dakota where we experienced an average daily high temperature of -10 °F (-23 °C). Before my trip, my daughter gave me a pair of very heavy warm wool socks to wear for returning to that cold climate. To me, this gesture from a child who missed her dad represents the concept of "Ollin." She and all of the other members of my family, and my partner's family, were making as much of a sacrifice as my own and those of my business partners to see the realization of a new company.

Rare Earth Elements: Significance in Modern Technology

The major business of our startup company focuses on the extraction and separation of rare earth elements from ores. The rare earth elements, also known as the lanthanides, are a group of 18 elements, 17 of which are naturally occurring (Figure 1). The rare earth elements (REE) are becoming ubiquitous in our daily lives, from the catalytic converter on cars, to the magnets in cell phone speakers they are everywhere (Figure 2) (*2*, *3*). The ready availability of the natural resources for the creation of chemical precursors is declining, while the consumption of those same natural resources is climbing (*4*). It is estimated that China currently produces 90-95% of the world's rare earth oxides and is the leading producer of samarium cobalt (SmCo) and neodymium iron boride (NdFeB) magnets commonly used in many electronic devices (*5*). China has clearly stated that they will no longer be exporting the rare earths as fine chemical precursors, due to their national demand being greater than the global supply (Figure 3) (*6*, *7*). U.S. trade representatives have stated that the quota imposed by China has caused, "…US manufacturers to pay as much as three times more than what their Chinese competitors pay for the exact same rare earths" (*8*). The shortage of suppliers for rare earth elements has also created vulnerabilities in US defense capabilities. The Government Accountability Office estimates that it could take 15 years or more to revamp the rare earth supply chain so that NdFeB magnets, a major component of defense electronics, could be produced domestically in the required quantities (4). Because of the serious potential consequences for both US industry and national defense capabilities created by waiting decades to have a sufficient domestic rare earth supply, it is evident that new technologies and methods are needed for producing REE in a much shorter time.

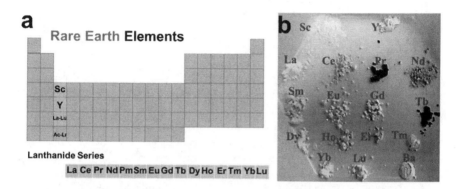

Figure 1. Rare earths and their oxides. (a) Locations of rare earth elements in a periodic table. (b) Photo of rare earth oxides refined from ores using the proprietary process developed by Rare Earth Salts, LLC.

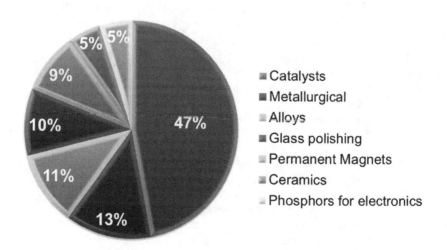

Figure 2. Global demand of rare earth elements in different technology sectors.

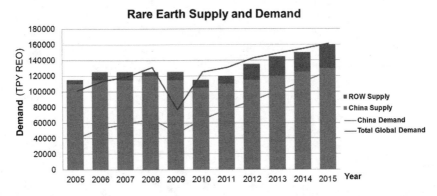

Figure 3. Supply and demand of rare earth elements: Past data and projections.

Refining operations are a bottleneck in the domestic production of rare earth materials and stopping ore mined across the US from being efficiently turned into useful products. Because refining operations associated with the mining industry have long been viewed as environmentally unfriendly, opening new foreign or domestic rare earth refineries using current technologies poses several potential problems.. Traditional methods for the synthesis of precursor fine chemicals from natural resources are wasteful in both the expenditure of energy, and in the relatively high concentrations of elements of interest which have been left behind in the mine dumps (*9, 10*). Because of these inherent problems, it is necessary to look for entirely new methods of refinement which are energy efficient, environmentally friendly, and can quickly be implemented on a large scale.

Several factors make purification of rare earths complicated (*11*). First, the 17 elements all tend to occur together in the same mineral deposits. Because they have similar chemical and physical properties, it is difficult to separate them from one another. Secondly, they tend to occur in deposits with radioactive elements, particularly thorium and uranium (*12*). Those elements can become a threat if the "tailings," the slushy waste product of the first step in separating rare earths from the rocks they are found in, are not dealt with properly. Finally, the chemistry of the rare earths is characterized by the similarity in the properties of their trivalent ions and their compounds. The structure, solubility and coordination chemistry of rare earth ions in solution along with the electronic configurations, size relationships and various oxidation states of the rare earths have all been reported and studied in hopes of finding better means of separations (*13–16*).

Rare Earth Extraction Needs "Quantum Jumps" More Than Incremental Advances: A "Rare" Opportunity

Sadly, it seems that all too often today business, and academic researchers choose the safe path of making small incremental advances to a product line or in an area of research rather than taking a chance on a wild idea. We chose to throw out the last 50 years of research on REE refining and start over. By starting with a clean slate, listing the desired outcomes and making environmental friendliness a top priority we were able to come up with a completely different approach to extracting the elements from the ore. Additionally, we found that ion exchange columns, which are commonly used for the separation of the REE from each other after extraction, could be replaced with less expensive off-the-shelf equipment and that the resulting byproducts were considered to be generally safe for the environment.

When approaching a complex problem like extraction of select elements from ore, it is vital to think outside the box. Just because it is commonly accepted that there is only one way to do something doesn't mean that is actually the case. Little has changed in the separations of the REE in more than 50 years. There have been incremental advances, but nothing significant. While considering the process it occurred that there was a completely novel way of separating the elements from each other that didn't involve ion-exchange columns and would still produce a pure product. This process is more economical than the current methodology while the waste stream is non-toxic. As exciting as this discovery was, we also developed a new method for extracting the elements from the ore at greatly reduced cost and improved efficiencies. Both of these processes together allow us to drastically lower the cost and environmental impact of the extraction of the REE. By deciding to throw out the conventional wisdom we were able to determine that incremental advances were simply not enough. Previous convention taught us that in order for an ore body to be economically viable it needed to be at least 8% rare earth oxides. Utilizing our new methods ore bodies of much lower concentrations can be viable. This allows us to greatly increase the known supplies of rare earths that are so valuable to our society.

While studying at the University of Nebraska-Lincoln we learned the value of serendipity (*17*). While investigating more effective ways to anneal cerium oxide catalysts, it was discovered that the pressure at which the catalyst was annealed played a large role in determining the overall activity of the catalyst. The large effect that low pressure annealing had on the catalytic activity of cerium oxide was not expected based on our initial hypothesis. However, as with many scientific advances, careful planning and execution in one area of our research led to increased understanding in a secondary area. After months of dedicated hard work aimed at understanding this new phenomenon, we were able to demonstrate that the catalytic activity of cerium oxide could be increased by lowering the processing pressure during post synthesis annealing. Although this was not the reason for performing the initial experiments, the benefit in our ability to prepare effective catalysts was greatly improved over what would have been expected had we pursued our initial research plan. Since becoming a full-time entrepreneur, I have discovered that the effect of serendipity cannot be understated. Each day an entrepreneur must decide whether the current path is the most profitable. The experience I gained in the lab evaluating new findings and deciding whether or not they merit moving in a new direction has proven to be an invaluable guiding light in this regard.

Combining out of the box thinking with serendipity is one path to success as an entrepreneur. As scientists we should always consider that there is more than one way to accomplish a task. With improvements in instrumentation and techniques, long held rules of thumb may no longer be valid. With nanotechnology and other advancements in materials increasing, it is likely to soon become the norm to mass produce exacting inorganic molecules that were, until very recently, considered to be novelties and of no particular use outside the laboratory. Now these compounds and technologies are helping us to rewrite what is possible and what is economical.

Initial Research and Funding

When you start any business you need cash. Initial research can be done out of pocket or via credit card debt; however, scientific research is expensive and soon a larger cash flow will become necessary. It is vital that before you accept any money, you obtain legal counsel. In the US money is highly regulated and if obtained incorrectly you may lose all rights to your company and everything you have created. Additionally an accountant or accounting firm who says no to all initial requests and proposals to spend money will keep you on the right path.

A valid business plan containing milestones and funding stages will provide you with a route to success. Additionally your business plan will help to reassure investors that you can turn your chemistry into profits. An excellent source for help writing a business plan is the local business college. While we, as chemists, have spent years perfecting the synthesis of matter; the students in the MBA programs have spent an equal amount of time learning to write a road map to success in the form of a business plan. Spending time with one of these students can offer insights to potential pitfalls that you can plan to avoid.

Angel investors can be either a boon or a burden to a new startup. There is a cost for their money and they will often require a significant portion of your company. Before approaching them, you and your partners need to meet with your attorney and accountant and determine how much you are willing to sell and what your bottom line is. When you meet with angel investors they have more to offer than just money; they have experience with startup companies, they have human resources in place, and they can provide benefits to your employees. Frequently they will want to help guide you through several milestones and will want to help put together a management team. It is in their best interest to see you succeed as that is the only way they will be rewarded with a return on their investment.

Another great source of startup and research money is government grants such as those through the NSF and SBIRs. If you are successful in being awarded one of these grants not only will that money be available to you, but you will be promoted as a leader in your field. The key to obtaining a grant is to have open communications with the grant administrator. The grant administrator is an often undervalued resource who is frequently willing and able to give advice on ways to strengthen your grant proposal. Additionally, if you have had good communication and are unsuccessful, the grant administrator can give you a review of your proposal and suggest changes for the next cycle.

When you start a chemical business you also need access to environmental health and safety (EHS) resources. For those working with academic or similar partners the EHS concerns are likely already in place. However for those who are starting in their garage you need an EHS partner who can advise you on safety, legal requirements for the workplace, and can provide proper disposal for used chemicals. Frequently the same company who can provide proper disposal can assist you in ensuring that your lab is safe and up to code. Additionally they can advise you if there comes a time when your research reaches a stage where your experiments need to be performed in more appropriate lab space. Once you outgrow your initial lab, finding a suitable work environment is the next step (Figure 4). Many cities, and universities, have industrial parks where you can rent or build a lab. Often in these industrial parks the EHS costs are included in the prices of the lab space.

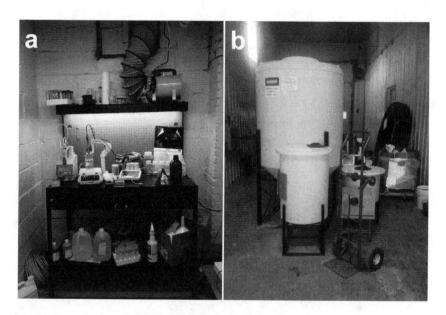

Figure 4. Progress of a startup company: From (a) a garage laboratory to (b) a scaled-up laboratory

Because of many reasons, a startup company wishing to purchase chemicals must be vetted by the chemical suppliers. The vetting process can take several weeks to months, and must be accounted for in your business plan. Much of the vetting process depends on how you have set up your company and what your state lists your business as. If, for example, your state lists you as a chemical company that will greatly speed up the process as opposed to if you are simply listed as industrial. It is in your best interest to have your attorney consider this when setting up your company as it is always easier to do it right the first time than to change it later. Again if you are associated with academia or a research institute, they will be able to greatly accelerate this process.

A well-designed business model to manufacture "products" is critical to the ultimate success of a startup company. In chemical industries, the "products" can be in different forms, from providing information such as in chemical analysis, writing chemistry software, to manufacturing chemical products. In the case of manufacturing chemical products, startup companies typically develop the intellectual products and associated synthetic schemes and seek established chemical manufactures to scale up the manufacturing of the products for sale in the market. However, in many cases such as mine, there are no facilities available which can adopt the proprietary chemical processes. Thus, it is necessary for the company to develop the scaling up process and convince the investors to invest in building pilot plants for economy of scale manufacturing.

One of the challenges for a chemist to realize is that there is a large difference between lab-scale, pilot-scale and industrial-scale reactions. Though chemists are taught in courses about the importance of chemical kinetics in terms of the transportation of mass and conduction heat in chemical reactions, scaling up the reactions from bench scale, grams-scale, 10s of kgs-per-day-scale, to eventually 1 ton-per-day-scale are daunting tasks. While these chemical engineering concepts are well discussed in college, to implement the applications of these concepts for new processes is complex. In my case, my company needs to develop custom-designed equipment and kg-scale process to extract and separate rare earths in oxide forms from ores. Rather than handling 100 mL or even 2-3 L beakers as for bench top scale, my company needs to handle liquids in the orders of 10 to 100s of liters per reaction. For a startup company in a similar situation, before even starting the actual R&D process, one would need to work with equipment supplier to design and purchase new equipment. Also, seeking legal advices from consulting firms on environmental and safety compliances regarding the planned processes is a must to avoid fines from EH&S. Working with chemical suppliers (in my case, ores suppliers) and their transportation issues are other necessary issues allow a "smooth" start before doing the "fun" experiments.

Another important concept in scaling up the reaction is the concept of thinking in "moles" than kilograms or tons. While this is a fundamental concept in high school chemistry, chemical industries often cite the production of chemicals in tons in the media and this is often used in the business language for chemical products. However, the conversions between "kgs and tons" to moles should be carefully considered when you make business decisions on the price of chemical products versus their production costs.

As you begin to grow you cannot hope to do all of the jobs yourself. How do you find individuals who are the right fit for a startup company? Everyone involved in a startup must wear many hats, and this is not for everyone. If you are advertising for people you should include in the job description that you are hiring for a startup company and that the successful employee will be very flexible in their duties. In early stages you are likely to be hiring people who will want to grow with the company. You are looking for people who will be dedicated and willing to stick with the company through the hard times. At these early stages, hiring people you already know can be a good work strategy. Also workers with former military service are frequently very flexible and dedicated individuals who are prepared to adapt as things change. It is reasonable for employees who join the company at an early stage to expect to grow with the company. They should be rewarded for struggling through with you in the beginning. But growth will come with the need for new individuals with specialized training. It is important that these individuals be carefully identified.

Business Partnerships

In the chemical industry you do not exist alone, even as a mineral refiner near the beginning of the production chain, we cannot hope to operate without some sort of strategic alliances with other businesses. As noted before setting up the groundwork with chemical suppliers can take a significant amount of time and needs to be initiated early on. Once you have a relationship with your supplier, get a personal account manager. As you grow the account managers can help you to prepare for delays and lag in supplies purchased in bulk that may not exist for those purchased at gram or Kg quantities. Also they can, and will, setup recurring shipments and assist you in finding the best deal on the grade of chemical you need.

As you scale up your relationship with the utility companies will change, your water and power needs will grow faster than your company. An account manager with the various utility companies becomes a necessity. If you are installing a large piece of equipment, a good relationship with the utility company can greatly speed the installation of the necessary resources to power that equipment.

As mentioned before an EHS contract service that can grow with you is vital to the success of any chemical company. Combined with transportation companies who have an excellent knowledge of interstate shipping requirements for chemicals will allow you to proceed as the quantities of chemicals required begin to take up entire trailers or box cars.

Each of these partnerships will assist you in growth and will help to eliminate the costs associated with doing business. Additionally they want you to succeed and they will assist you in planning and growth of your company.

Looking to the Future: Opportunities and Challenges

After you begin to grow, you realize that you must have a plan for the future. The old adage "if you fail to plan, you plan to fail" has never been so true as when you are in the initial stages of building a corporation. Keeping the end goal in mind and following your business plan even while it grows and changes is vital. Don't forget your partners, your families, your investors and those who have helped and supported you through the initial stages. Keep in mind that it is likely that at some point you will need to step back from the helm and focus on what you do best. If you are a chemist who loves innovation, step back and lead a group of hand chosen individuals who can help you develop new innovations. If you discover that you love business, consider allowing someone else to take over the daily operations and become the chair of the board.

Each business is unique; as you grow you will face challenges and uncertainties. You may discover that in order to do what you set out to do, that you need to assist others to start companies that will support yours. Every challenge that is overcome brings with it an opportunity to grow and learn. Staying strong and overcoming each challenge makes you more interesting and better prepared for the next time.

Each challenge, no matter how much of a crisis it appears to be initially, needs to be viewed as an opportunity. When treated in this manner the growth of the company will be ensured; because there will always be a new challenge that awaits you. Encourage everyone within the company to be innovative and to look for new ways to overcome and grow from daily challenges. Then every challenge becomes an opportunity and growth and success become more assured.

For those who are beginning your adventure as entrepreneurs: No matter what you do, you have worked hard, you have found a passion and you are the lucky few who can tell others that you choose to do what you do because you love it.

References

1. Hall, K. *Aspire : discovering your purpose through the power of words*, 1st William Morrow hardcover ed.; William Morrow: New York, 2010.
2. U. S. Geological Survey. *Mineral Commodity Summeries 2011*; U.S. Department of the Interior; U.S. Geological Survey: 2011.
3. Trovarelli, A. *Catalysis by Ceria and Related Materials*, 1st ed.; Imperial College Press: London, 2002.
4. Gschneidner, K. Rare Talents. *Technol. Rev.* **2011**, *114* (3), 13.
5. Grasso, V. B. *Rare Earth Elements in National Defense: Background, Oversight Issues, and Options for Congress*; Congressional Research Service: 2013.
6. Kingsnorth, D. J. Meeting the Challenges of Supply this Decade. http://files.eesi.org/kingsnorth_031111.pdf (accessed March 2014).
7. Bradsher, K.; Tabuchi, H. China Said to Widen Its Embargo of Minerals. *The New York Times*; Oct. 19, 2010.
8. Miles, T.; Hughes, K. China loses trade dispute over rare earth exports. *Reuters*; 2014.
9. Li, L.; Xu, S.; Ju, Z.; Wu, F. Recovery of Ni, Co and rare earths from spent Ni-metal hydride batteries and preparation of spherical $Ni(OH)_2$. *Hydrometallurgy* **2009**, *100* (1–2), 41–46.
10. Nash, K. L.; Choppin, G. R. *Separations of f elements*; Plenum Press: New York, 1995.
11. Gupta, C. K.; Krishnamurthy, N. *Extractive Metallurgy of Rare Earths*; CRC Press: Boca Raton, FL, 2005.
12. Cai, R.; Li, W. Separation of thorium from rare earth metals. *Beijing Ligong Daxue Xuebao* **1995**, *15*, 289–293.
13. Eriksen, D. Extraction of rare earth metals from carbonate-containing ores by water treatment with carbon dioxide injection; GB2464733A, 2010.
14. Gurov, V. A. Ecological method of complex extraction of nonferrous, rare and precious metals from ores and materials; RU2352650C2, 2009.
15. Plaksin, I. N.; Barysheva, K. F.; Astaf'eva, A. V. Study of rare earth metals by extraction. *Izv. Akad. Nauk SSSR, Otd. Tekh. Nauk, Metall. Topl.* **1962**, 185–191.
16. Yang, X.; Chen, D.; Liu, B.; Lu, N.; Ouyang, H.; Zhou, J.; Peng, S.; Lai, L. Method for online saponification and calcium ion removal of extractant for rare earth separation using calcium oxide, calcium hydroxide or calcium carbonate; CN102041383A, 2011.

17. Lawrence, N. J.; Brewer, J. R.; Wang, L.; Wu, T.-S.; Wells-Kingsbury, J.; Ihrig, M. M.; Wang, G.; Soo, Y.-L.; Mei, W.-N.; Cheung, C. L. Defect Engineering in Cubic Cerium Oxide Nanostructures for Catalytic Oxidation. *Nano Lett.* **2011**, 2666–2671.

Chapter 18

Technical Entrepreneurship Serving Industry: A Personal Story

Sharon V. Vercellotti* and John R. Vercellotti

V-LABS, INC., 423 N. Theard Street, Covington, Louisiana 70433
*E-mail: v-labs@v-labs.com

As the job market has tightened, more professionals are considering entrepreneurship as an alternative career. Entrepreneurship built around unique personal talent and experience can convey valuable service to industry. Universities are offering entrepreneurship courses for non-business students. Starting a small chemical business can be a challenge. The authors review their experience starting a carbohydrate laboratory in Louisiana and discuss some of the lessons learned in the process. Sharing these ideas is often valuable to those considering taking similar steps. The current status of small business in the United States and the globe will be reviewed as well, focusing on innovative scientific businesses.

Introduction

The North American Industry Classification System (NAICS) assigns category numbers to each of the manufacturing sectors in industrial businesses. The chemicals and allied products manufacturing sector is labeled NAICS 325. The complete economic census of these industries is made every five years. The most recent data are from the 2011 U.S. Economic Census. The previous census for comparison was in 2006 (*1*). Through these valuable data, economic balance and direction can be tracked for the chemical manufacturing industry. A small business is defined as one having 500 employees or less. For sake of comparison in 2006 there were 13,476 such chemical establishments. In 2006 25.4% of all chemical businesses were large businesses. Although 74.6% were small businesses, actually 46.2% of these companies had fewer than twenty employees (*1*). The US Economic Census covers a broad spectrum of information, besides

© 2014 American Chemical Society

the numbers of firms and establishments. The Census also collects data on the number of employees and the annual payroll by employment size of the enterprise for the United States. From these data in 2006 about the total number of employees, The Census estimated that 69.8% of all employees in the chemical industries were working for large firms having more than 500 workers (*1*).

The next NAICS 325 data for comparison are from 2011 because the cross section of chemical manufacturing data is compiled only every five years by the US Economic Census. The total number of chemical manufacturing businesses with NAICS 325 had dropped to 12,914, or chemical manufacturing, NAICS 325, had reduced in size approximately 4.1% (*2*). At the same time the number of small chemical businesses rose while the large chemical companies with greater than 500 employees cut back their domestic production, research, and development staffs, reflecting a totally smaller chemical work force (*2*). The 2011 US Economic chemical manufacturing census reflects this shift in the number of establishments versus the number of employees as 80% of the 675,000 employees being in the small chemical businesses category and 20% in the larger firms employment. Such is a significant shift in the traditional employment pattern of professionals in all sectors of the chemical industries (*2*). Chemical employment shifts were also accurately reported by McCoy et al. in the C & EN in 2010 (*3*).

As an example of the changes in chemical industry employment patterns on October 1, 2013, Merck Pharmaceutical Co. announced that it was laying off 20% of its global work force of 81,000 workers, or approximately 16,000 employees (*4*). Where these workers will be assimilated is an open question, but many must seek niches in small business since there are fewer openings available in the large industries. Putting this into the perspective of the extent of all small businesses in the United States during the 2006 to 2011 period, there were 6.02 million firms in the United States with 5.37 million firms employing less than 20 people. There were 18,071 firms or 0.3% which employ more than 500 people. More than 99% of firms in the United States are small businesses (*2*).

Economic Significance of Job Creation in the U.S.

According to Stephen J. Davis and coauthors, who have studied the question of job creation in the U.S. industrial economy for the Kauffman Foundation, "More than one-third of job creation is due to the entry of new businesses" (*5*). Other supporting evidence from the Kauffman Foundation shows that, "From 1980-2005, firms less than five years old accounted for all net job growth" (*5*). In the perspective of innovative business employment, U.S. small businesses employ more scientists and engineers than large businesses. The ratio of small to large business for these technically skilled employees is 32 percent to 27 percent.

The number of scientists and engineers in small businesses is even greater than the total of this category in all the universities and federal labs combined at 32 percent in the small businesses compared to 29 percent in those dedicated research institutions (*6, 7*). Much innovative productivity comes out of small technical businesses as reflected in the number of patents issued to them per dollar of research and development. Thus the small technical businesses are more productive than the large businesses by this comparison with a ratio of 5 to 1. Compared to the universities the entrepreneurial small technical businesses are more productive using this intellectual property standard by a ratio of 20 to1. At the same time Federal research funding distribution is hardly equitable for the small technological businesses. Larger firms receive 50.3% of Federal research funding, universities, 35.3%, and small businesses 4.3% (6).

Examples of Funding Sources for Small Science or Engineering Businesses

The success of a small scientific business depends on many factors such as realistic marketing research for the product or service to be offered as well as sourcing for funding to support it. There are many routes that a practicing chemist as an entrepreneur can follow. The unique problem solving skills and familiarity with technology that chemists have acquired at all degree levels permit them to be successful in a number of areas. These include the usual laboratory analysis or syntheses of innovative products, chemical information management, patent law, technical editing, environmental management, fine chemicals marketing, pharmaceutical regulatory oversight, and specialized technical publishing.

The startup company needs incorporation as well as liability safeguards through insurance. These costs can sometimes be borne by capital savings or family and friends' interest in investment for the idea. More often the initial choice of financing is through a well-documented bank loan with collateral in the form of contracts or existing property. Even modest laboratory facilities for chemistry and chemical product manufacturing represent a significant investment. Over and above just building and equipping the technical site to carry out the work, many other investments in environmental safeguards, utilities, and facility upkeep as overhead must be taken into account.

Venture Capital as a Significant Source of Funding for Entrepreneurs

Investment in new businesses always requires careful examination of the foundations on which the business is established. Richard Buckminster Fuller recommended looking carefully at the structure of his new buildings and suggested, "You never change things by fighting the existing reality. To change something, build a new model that makes the existing model obsolete." (8). This is also the model of a new foundling business.

Depending on the end product, the cost-effectiveness of successfully achieving that goal even at maximum risk must come down as the marketable value of the product rises. Ideally venture capitalists support a worthwhile idea because they recognize value in the new firm's organization and product goals. The product should have winning characteristics for the venture capitalist to risk his money as a stakeholder.. The entrepreneur needs to be open to being a partner while the venture capitalist must bring knowledge of the real business world and how to motivate a company culture for the entrepreneur.

A recent PricewaterhouseCoopers report (9) noted that of one year's $17.68 billion dollars, $1.68 billion was invested in startup/seed companies (9.33%); $4.63 billion went into early stage companies (26.18%); $5.84 billion into expansion companies (33.1%); and $5.92 billion into later stage companies (33.5%). Many of these investments are in small chemical or pharmaceutical and medical device companies. The principal distributions are in biotechnology 20%, software 17.5%, and medical devices 14.2% (9).

The United States Small Business Administration

For more than sixty years since its establishment under President Eisenhower, the dedication of the United States Small Business Administration(SBA) has been "to maintain and strengthen the nation's economy by enabling the establishment and viability of small businesses and by assisting in the economic recovery of communities after disasters" (10). The agency does this through their traditional "3C's" of capital, contracts, and counseling. In addition to guaranteeing bank loans to entrepreneurs at present up to 90% of the loan to a bank or other lending agency in the event of default, the SBA helps lead the Federal Government's efforts to deliver 23% of prime federal contracts to small businesses.

Established small businesses engaged in innovative development of new products are eligible for competing applications to funding through the Small Business Innovation Research (SBIR) program. This program was begun during the Reagan Administration in 1982 under the program founder, Roland Tibbetts, whose intentions were, "to provide funding for some of the best early-stage ideas – ideas that, however, promising, are still too high risk for private investors, including venture capital firms". The program is coordinated by the U.S. Small Business Administration in which 2.5% of the total extramural research budgets of all federal agencies with extramural research budgets in excess of $100 million are reserved for contracts or grants to small businesses. In recent years that has amounted to over $1 billion dollars a year to small technical businesses in research funds. These objectives are listed in highlighted emphases (10):

- Use small businesses to stimulate technological innovation.
- Strengthen the role of small business to meet Federal research and development needs.
- Increase small business participation in Federal research and development.

The Small Business Innovation Research program provides these budget Federal dollars.

- Current posting of some fourteen Federal agencies and departments with 2.5% of extramural research budget dedicated to the SBIR program.
- Phase I grants. $150,000 for six months.
- Phase II grants. Up to $1,000,000 for two years.

As of recent data compilation the SBIR has generated some 87,000 U.S. and world patents. Fifty percent of the products from these projects have realized commercialization for firms that have received Phase II grants. Twenty-five percent of research and development "R&D 100 Awards" were given for the one hundred most significant new technology products each year that have been funded by SBIR. The related small business-academic research program called the Small Business Technology Transfer Program (STTR) uses a similar approach to SBIR to expand public/private sector partnerships between small businesses and nonprofit U.S. research institutions. The STTR program is funded at 0.3% of the agencies' extramural research budgets. In recent years this has amounted to over $100 million (*11*).

Foundations and Consulting Organizations To Assist Entrepreneurs

- **Kauffman Foundation** for research, education, technical assistance and policy. This foundation dedicated to expanding the entrepreneurial enterprises of America has many services available: FastTrac^{R}NewVenture™ aiding aspiring-early stage entrepreneurs and FastTrac^{R}GrowthVenture™ helping existing entrepreneurs improve their business (*12*).
- **Goldman Sachs *10,000 Small Businesses*** is part of The **Goldman Sachs** Group, Inc. as a leading global investment banking, securities and investment management firm that provides a wide range of financial services to entrepreneurs. This fund represents a $500 million investment of the Group. Its goal is to help entrepreneurs create jobs as an economic opportunity at no cost to the entrepreneur. This assistance from Goldman Sachs provides education, financial capital, and business support services to small business start-ups (*13*).
- **NineSigma, Inc.**, provides open innovation (OI) services that connect clients with companies, researchers and individuals worldwide. These services include collaboration strategies and tools which assist in technology searches, project selection and consulting with innovative leaders and their teams (*14*).
- **National Collegiate Inventors and Innovators Alliance** promotes invention, innovation, and entrepreneurship at the college level. In addition to collaborative experiential learning programs, encouragement and formal teaching programs are made available to the new generation of innovators and entrepreneurs (*15*).
- **InnoCentive** is a powerful online forum from which major companies outsource scientific innovation through financial incentives. Contracts are made to award winning proposals for solving difficult industrial problems and let from major U.S. companies, government, and other non-profits. The motto of InnoCentive is, "Be a solver" (*16*).

V-LABS, INC., as a Model of a Small Chemical Business Startup

In 1979 the coauthors, as two practicing chemists and entrepreneurs, wished to provide in-house laboratory analyses, specialized glycobiology model compounds, and consultation to carbohydrate based industries in a small chemical business. As a laboratory dedicated to the chemistry and biochemistry of the carbohydrates, we aimed to serve a wide cross section of the biomass conversion, sugar, nutraceuticals production, carbohydrate biomedical research, and carbohydrate-based materials industries.

To perform the needed services for these industries on an economically sustaining level for our company we needed to perform careful cost accounting on every step of the work performed. The marketing of scientific services is highly competitive, but is the real world into which business professionals in chemistry must proceed. In addition to services we also envisaged a catalog of purified carbohydrate model compounds for analytical standards as well as for use in biological metabolic research which we now manage with a partner from the United Kingdom, Dextra Laboratories.

In order to achieve these goals we had to obtain support from several partnering sources.

- Committed contractual work: 2 years
- SBA guaranteed bank loan plus personal savings, contracts:
- No salary for the first 2 years
- Today, with inflation index our startup would require 3.2 times the money as then.
- Prime rate interest on business loans in 1979 at the end of the Carter era was 21%/per annum which the SBA secured for us at 11.2% over ten years.

From the very inception of the business as part of the overhead of operation an attorney was made a committed counselor for all aspects of the business. The many legalities involved in establishing ownership and legal responsibility were spelled out to us with the assistance of our attorney. In addition to incorporating the company with the Secretary of State of Louisiana as a Subchapter S company, there were a number of questions such as whether the zoning on the land where we wished to build permitted laboratory operations. Such questions involved emissions from our fume hoods and wastewater drainage into the sanitary sewer system.

Overhead to operate the facility has included careful monitoring of quarterly Federal and State taxes, salaries, utilities, liability and property insurance. In addition to the attorney the next most important consultant has been our accountant who also knows the ins and outs of the tax codes as well as investment advantages that we would have as a small incorporated business. Adjusting cash flow to meet payroll and social security payments are also under the eye of the accountant. In addition to employee benefits, state workman's compensation, liability insurance for the facility, and funds to provide adequate 401K accounts or similar set-asides for service need to be planned as the business grows, A full service bank is absolutely necessary to provide both fast invoice transfers or receivables payment by wire. Security of account codes and protection from fraud carried out by many devious approaches such as computer code hacking of our passwords or pin numbers must be monitored by the bank. V-LABS has been the victim of two large overseas wire frauds which fortunately our bank was able to screen and detect without actual loss of funds. Establishing customer credit card purchasing from us as well as our use of the most economical forms of charge card services has been on-going since the beginning of the business.

The planning of the building with secure safety features also required estimates of maintenance on the property. The design of our laboratory building necessarily had to involve resistance to corrosion, sewer, water, gas, and electrical conduits which were up to code for industrial use. Our building was designed for maximum efficiency in daily work patterns as well as climate control and air exchange for occupational health and safety. Emergency water showers had to be installed as well as fire extinguishers at convenient places throughout. At the time we were beginning the building and the laboratory business the Toxic Substances Control Act of 1976 was already in force. We had to anticipate all the Environmental Protection Agency regulations on manufacturing as well as detailed records about chemicals entering the laboratory as well as their ultimate disposal ("cradle to the grave" responsibility of the owners). As part of our fiscal planning we also had to anticipate costs of certified disposal of all chemicals through setting aside enough funds to cover these costs.

All of this had to be done before we even started digging a foundation and laying down underground plumbing. The building occupies 1600 square feet, which has been more than adequate for our work space. Utilities such as telephone, gas, electric, sewer and water services were budgeted and ordered installed. Air handling equipment in addition to fume hoods required installation of high volume air exhaust fans were built into the building and safety features such as fire and overhead safety showers built. Costs of electricity for climate control in the building are considerable both for heating and cooling.

Modern instrumentation has been maintained through the years permitting us to solve a variety of analytical and synthetic problems. Our analytical chemistry tools have always provided adequate information to solve our customers' problems. We have depended upon a useful cross-section of well maintained and serviceable instruments such as gas and liquid chromatographs with computer integrating softwares, diode array ultraviolet-visible spectrometer, infra red instrument, vacuum pumps, ovens, preparative centrifuges, ultra- and nanofiltration for polysaccharide isolation, physical measurement instruments such as electronic refractometer or solid state pH and conductivity meters, These instruments carry our work load but now are much out of date for business expansion. In the modern world of carbohydrate chemistry and glycobiology liquid chromatography is done with high performance ion chromatography with pulsed amperometric detection and autoinjector for multiple continuous sampling. The bench top gas or lilquid chromatograph-mass spectrometer also with autoinjector and computerized libraries for compound identification are today routine for accuracy as well as rapid and large volume throughput. In much of our preparative work we would be well served with a Fourier transform infra red spectrometer that would permit us to identify a wide variety of samples with spectral library search capabililty. Although our ultraviolet and refractive index detectors on the liquid chromatograph permit us to perform routine size exclusion chromatography on soluble polysaccharide samples we would find advantage in having a viscosimetric and laser light scattering detector to obtain more physical definition of polymer samples. The overhead on instruments such as nuclear magnetic resonance or large mass spectrometers would be prohibitive. We have had arrangements for per sample analyses with reliable laboratories to have these highly specialized analyses run.

With the growth of computer and on-line communication in every form we have built an efficient in-house computer network within our building to handle correspondence and data storage on large storage drives. With daily overseas business to transact orders, we also need fast access data handling and have built up these capabilities over the last thirty years since installing our first computer systems. The many software packages that must be constantly updated and renewed have also made possible a lot of our graphics and report format utilities. Costs for high speed data transmission through the telephone system are expensive and must be budgeted each year.

One of the greatest assets we had for promoting our business has been the associates we have had for networking through the American Chemical Society in the Small Chemical Businesses Division as well as the Divisions of Carbohydrate Chemistry, Cellulose and Renewable Materials, and Agricultural and Food Chemistry. Not only have we had uninterrupted continuing education through these Divisions as world class resources, but we have been able to partner with members to join into projects and problem solving which we would not have been able to do on our own. By dividing up a difficult project each of the cooperators is able to give their best and the end result is a well solved problem for the client.

Although there are only four of us working at V-LABS we are able to manage prompt and effective technical support when needed. In local communities many opportunities exist for potential young scientists to gain expertise and broad knowledge in science. Innovative small businesses can engage students in experiential STEM education on the local level by providing part time intern positions for high school and college level students. These opportunities may contribute to better understanding of potential careers in the sciences by expanding views from the classroom. We have recruited some 25 students from local high schools and a nearby college to work part time in our laboratory since our founding thirty-four years ago. We have also served as mentors in local high school internship programs. For our business the value of this student help is very supportive of our production and end product value. In our laboratories the students gain first-hand practice in effecting chemical analyses or preparative syntheses. By buying into the cost-estimation and profitability of their work the students also come to connect the real-world value of their team efforts in the light of the academic lessons they have received in the classroom.

During the 1980's and 1990's V-LABS, INC., was fortunate enough to be awarded three U.S. Small Business Innovation Research projects. Two of the contracts were from the National Institutes of Health and one, from the National

Science Foundation. Although we had successful Phase I periods on these projects we had to move on to other developments in the company and did not pursue Phase II projects. We do encourage any small business entrepreneurs in the audience to follow up leads in the Federal SBIR programs. There have been many interesting and challenging projects over the years, literally hundreds of them, The whole thirty-four year experience in the business of affording chemistry on a wide variety of carbohydrate related projects is that there has never been an uninteresting, trivial, or boring experience in all those years. In these later years we also have had to plan how the business would be either morphed into another type of operation, sold, or closed altogether. Since we are still actively involved in much commerce and laboratory service none of these exit strategies are being pursued.

Conclusions

Opportunities will continue to expand small chemical businesses even as large industrial organizations in the pharmaceutical, chemical materials, and food sectors become less economically attractive. In small chemical businesses such as are served by the Divisions of the American Chemical Society there is a synergy of talent and resources. From this comes beneficial alliances and a competitive advantage for these small chemical businesses to thrive. In the overall employment market for professionals in chemistry, new opportunities arise for application of the broad highly disciplined skills of these individuals. On a national and global scale there will be a greater equilibrium of capital and talent to reinforce these economies.

References

1. U.S. Economic Census. Number of firms, Number of Establishments, Employment, and Annual Payroll by Employment Size of the Enterprise for the United States, All Industries; 2006.
2. U.S. Economic Census. Number of firms, Number of Establishments, Employment, and Annual Payroll by Employment Size of the Enterprise for the United States, All Industries; 2011.
3. McCoy, M.; Reich, M. S.; Tullo, A. H.; Tremblay, J.-F.; Voith, M. Industry slashes thousands of jobs. *Chem. Eng. News* **2010**, *88* (27), 50–53.
4. Thomas, K. Merck plans to lay off another 8,500. *The New York Times* **2013** (October 2).
5. Davis S. J.; Haltiwanger, J.; Jarmin, R. *Turmoil and Growth: Young businesses, economic churning, and productivity gains*; Kauffman Foundation, June, 2008. URL http://sites.Kauffman.org/pdf/TurmoilandGrowth 060208.pdf.
6. *Why are high-tech small businesses so important to the United States?*; Small Business Technology Council, 1156 15th Street NW, Suite 1100, Washington, DC 20005; 2008.

7. Frequently quoted from R.B. Fuller and cited in *Goodreads Quotes*. URL http://www.goodreads.com/quotes/13119-you-never-change-things-by-fighting-the-existing-reality-to.
8. Evans, K. The quiet power behind U.S. job growth. *The Wall Street Journal* **2011** (November 19).
9. PriceWaterhouseCoopers; 2012. URL http://www.pwc.com/us/en/press-releases/2012/venture-capital-investments-q3-2012-press-release.jhtml.
10. SBA charter. The SBA was created on July 30, 1953, by President Eisenhower with the signing of the **Small Business Act**, currently codified at 15 U.S.C.ch.14A. The Small Business Act was originally enacted as the "Small Business Act of 1953" in Title II (67 Stat. 232) of Pub.L. 83–163 (ch. 282, 67 Stat.230, July 30, 1953).
11. URL http://www.sba.gov/category/navigation-structure/about-sba. Small Business Jobs Act of 2010. URL http://www.gpo.gov/fdsys/pkg/PLAW-111publ240/pdf/PLAW-111publ240.pdf.
12. Kauffman Foundation. URL http://www.kauffman.org/.
13. Goldman Sachs. URL http://www.goldmansachs.com/citizenship/10000-small-businesses/US/.
14. NineSigma. URL http://www.ninesigma.com/.
15. National Collegiate Inventors and Innovators Alliance. URL http://nciia.org/.
16. InnoCentive. URL http://www.innocentive.com/.

Diversity and Inclusion

Chapter 19

A Top-Down Approach for Diversity and Inclusion in Chemistry Departments

Rigoberto Hernandez[*] and Shannon Watt[†]

School of Chemistry and Biochemistry, Georgia Institute of Technology, Atlanta, Georgia 30332-0400
[*]E-mail: hernandez@gatech.edu
[†]E-mail: shannon.watt@gatech.edu

The Open Chemistry Collaborative in Diversity Equity (OXIDE) works together with chairs and thought agents in the leading research-intensive chemistry departments to advance diversity systemically and sustainably. Through discussion with social scientists and evaluation of our discipline, we are working to identify effective practices, policies and procedures that can be implemented within the chemical academy to advance diversity equity. Within this effort, diversity is interpreted broadly; it includes gender, gender identity, race-ethnicity, disability status and sexual orientation. OXIDE organizes the biennial National Diversity Equity Workshops (NDEWs), held most recently in April 2013. NDEW focus topics vary and have included sessions on contributing factors; interventions; organizational structure, behavior and dynamics; and recruiting, hiring, retention and promotion. NDEW2013 was the first of the chemistry-directed diversity equity workshops to include a specific focus on lesbian, gay, bisexual, transgender, queer, intersex and questioning (LGBTQIQ) equity. OXIDE gathers annual faculty demographics data from over 100 research-intensive chemistry departments and partners with *Chemical & Engineering News (C&EN)* to publish the results. OXIDE also maintains a dialogue within the collaborative through surveys, print and web dissemination, campus visits and other avenues. In this chapter, we argue why leading research-intensive chemistry departments must advance

© 2014 American Chemical Society

diversity equity within their faculty ranks and discuss how OXIDE catalyzes this transformation.

Introduction

The Open Chemistry Collaborative in Diversity Equity (OXIDE) was founded in 2010. It grew from the 2009-2010 Diversity Equity Workshop Planning Committee (DEWPC). The latter was initially formed to provide continuity to the prior workshops—funded by the National Science Foundation (NSF), National Institutes of Health (NIH) and Department of Energy (DOE)—held to address the alarmingly low representation in the leading research-intensive chemistry departments with respect to women (*1*), racial and ethnic minorities (*2*), and chemists with disabilities (*3*). These early workshops shared the hypothesis that substantive progress could be made only if the leadership of the leading research-intensive chemistry departments was engaged in changing the culture through dialogue, with social scientists providing perspective and solutions backed by solid research. Despite the successes of these workshops in changing policies and procedures to advance a more diverse chemical workforce, they lacked continuity because they were ad hoc events without supporting infrastructure or the promise of long-term follow-up. In recognition that each department has only one department head or chair to attend such events annually or biennially, OXIDE introduced the National Diversity Equity Workshops (NDEWs) based on the notion that all issues related to diversity—whether in common or particular to a given group—should be discussed within a single conversation between chairs, social scientists and representatives of diversity communities. In addition to gender, race-ethnicity and disability status, OXIDE has also included sexual orientation and gender identity as another critical component of diversity that must be addressed within chemistry departments. NDEWs were held in 2011 and 2013. They are planned to continue biennially until at least 2017.

OXIDE's partnership model (see Fig. 1) maintains continuity between NDEWs. OXIDE works with *Chemical & Engineering News (C&EN)* to publicize the faculty demographics obtained in cooperation with chemistry departments; gathers information from social scientists and diversity communities about barriers and solutions to diversity equity; disseminates the information among all the partners; and consults with individual chairs and departments on implementation of diversity-related initiatives.

A critical component of the OXIDE program is the recognition that substantive and sustained change in the climate of chemistry departments must arise both top-down and bottom-up. The bottom-up approach entails training and encouraging new scientists from diverse groups to pursue academic positions, and this model has been adopted by many successful training and mentoring programs. The top-down approach entails changing the culture, policies and practices of existing chemistry departments and universities so as to make academic positions attractive to under-represented scientists as a career choice. This can only be achieved in partnership with the leadership of chemistry departments, but requires active engagement with outside groups such as OXIDE to enable coordination across the academic community.

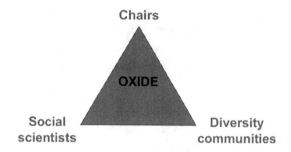

Figure 1. A graphical representation of OXIDE's partnership model connecting department chairs, social scientists and diversity communities. OXIDE serves as an active conduit of information between all three groups.

Diversity Defined

When speaking about diversity, there is invariably a question as to what exactly the term means, and what constitutes a diverse group. We interpret diversity to mean the inclusion of the other, where "other" is anyone unlike oneself. The other can differ in any of a number of attributes from categories such as gender, gender identity, race, ethnicity, disability status, sexual orientation, socioeconomic status, culture, life experiences, ideas, geography, university pedigrees, political ideology, country of origin (international), and others. A group is therefore diverse if its membership includes individuals expressing different attributes and which does not actively or passively exclude individuals with particular attributes.

It is ironic that the inclusion of people from some categories (such as country of origin) has historically been used as an indicator of a world-class department, but that the presence or absence of others (such as domestically-educated first-generation immigrants) has been ignored entirely in such rankings. Presumably the strongest departments are those that draw from the broadest possible pools of individuals. An international presence, for example, within a faculty indicates that the talent pool was broadened to include a larger—that is, worldwide—population. Consequently, the National Research Council (NRC) and others use counts of international faculty and students (4) as part of their rankings criteria. By analogy, the absence of under-represented groups should indicate that the pool was actually smaller, and be an indicator of diminished quality. Interestingly, the recent NRC rankings study (4) found little correlation between rankings and the level of participation from under-represented groups within faculties. Whether this finding is due to biases in the academic system, statistical insignificance arising from the very small number of professors from particular under-represented groups, or other factors is not known. However, the fact remains that there is a dramatic difference in the demographics of the faculty in comparison with the general population, as summarized below. This disparity suggests that we are not drawing the best talent equitably from within this nation's population and that our failure to do so is incompatible with the academic driver towards excellence that relies on attracting and retaining the best talent from the greatest possible pool of individuals.

In this chapter, as in the OXIDE effort, we focus on four diversity categories—gender and gender identity; race and ethnicity; disability status; and sexual orientation—because available data indicate that members of these groups are under-represented in the academic chemistry workforce in comparison to the demographics of the broader U.S. population. The data discussed below clearly indicate the existence of under-representation with respect to the participation of women and members of racial and ethnic minority groups. Unfortunately, the numbers are so small—and often lessened by underreporting—with respect to several of the categories that the level of under-representation is hard to measure. Indeed, the relative absence of said individuals in any given scientific gathering is palpable. The mismatch in the demographics between the broader population and the faculties is inequitable. The drivers that lead to it are the diversity inequities which OXIDE seeks to eliminate in academic chemistry departments by encouraging them to make doing so part of their excellence mission.

The Broader Context

Society as a whole has yet to truly remove the diversity inequities faced by individuals belonging to under-represented groups. Indeed, the lack of diversity equity in the workplace throughout all sectors, not just the chemical workforce, is presently receiving much attention from the public. Despite the many advances that women have made in attaining executive positions of companies and universities, many barriers to equitable participation continue to exist. For example, recent publications by Susan Patton (5), Anne-Marie Slaughter (6) and Sheryl Sandberg (7) describe some of the existing challenges—that is, diversity inequities—that women continue to face in balancing their private and professional lives in comparison to their male counterparts (8). Underrepresented racial and ethnic minorities are occupying positions at the top of the business and political world—including current notable examples such as Barack Obama (President, USA), Don Thompson (CEO, McDonald's), and Sonia Sotomayor (Associate Justice, Supreme Court)—and yet their representation is still far from equitable across the middle and upper tiers of leadership in the public or private sectors (9). The barriers faced by individuals whose real or perceived sexual orientation, gender identity or gender expression lies in the lesbian, gay, bisexual, transgender, queer, intersex, and questioning (LGBTQIQ) spectrum remain clear through current newspaper accounts and headlines devoted to not-so-prominent athletes who reveal their gender identity or sexual orientation, and in the intensity of the battle for marriage equality. Moreover, there is no doubt that great LGBTQIQ scientists such as Alan Turing (10) suffered dramatically because of the prejudice they faced, and that some present-day scientists remain hesitant to fully reveal their gender identity or sexual orientation. The aging of America, as well as the return of wounded American soldiers from military action, has shone a spotlight on acquired disabilities and the obstacles generally faced by individuals with disabilities in the workforce. While these examples point to the difficulty of the resolving existing diversity inequities, they also create an opportunity for academia to help in identifying solutions and leading the rest of the nation forward if we are able to resolve these issues within our own ranks.

The Rationale for Diversity

Universities and their chemistry departments, in particular, are driven by metrics indicating quality with respect to their mission to advance science and train students. Invariably, the metrics involve some kind of interpretation, making them difficult to quantify absolutely. In the business world, it is simpler to quantify success (through profits, for example), though it may be hard to ascertain the impact of any one individual in an absolute sense. Nevertheless, the so-called business case for diversity is a hypothesis that has been put forward (*11*). It posits that through greater diversity and inclusion, a business unit can gather a wider range of perspectives and knowledge, thereby resulting in enhanced decisions and actions that cater to a broader customer base in the increasingly global marketplace. Furthermore, when employees feel that they have a more diverse and inclusive workplace, personnel turnover is reduced, and employees tend to be more engaged and productive. While the evidence for the business case is still emerging, there do exist reports supporting the overall hypothesis that the return on investment for companies is higher when the working group is diverse (*11–14*). Equally importantly, having a reputation as an accommodating work environment—that is, one that is necessarily receptive to diverse employees—allows a company to draw from a greater workforce pool, thereby facilitating the hiring of the most talented individuals.

The business case for diversity is also relevant to chemistry departments, to the extent that each is a business unit. Such departments can likewise benefit from greater diversity (*15*). The academic workforce consists of talented and successful scientists of every gender identity, race-ethnicity, disability status, and sexual orientation. Those departments that are able to successfully attract and retain diverse talent are therefore at a competitive advantage. In addition, the combination of diverse faculty and more equitable climate in a given chemistry department advances the department's educational mission and presumably enables it to attract more diverse cohorts at all levels. Evidence of the latter may be illustrated in the successes of the Louisiana State University Chemistry Department in attracting and training a comparatively large number of under-represented minority doctorates (*16, 17*).

Diversity Demographics

Substantial mismatches clearly exist between the demographics of the U.S. population and those of the academic chemistry community at nearly all post-secondary educational and faculty levels. This remains true even though the demographic groups on which OXIDE focuses—women, racial and ethnic minorities, individuals with disabilities, and individuals who identify as LGBTQIQ—comprise a significant proportion of our nation's people (*18*). The great disparity in gender, racial and ethnic demographics was highlighted early on by Donna Nelson through her diversity surveys (*19, 20*). She used the NSF ranking of chemistry departments according to chemical expenditures, and this practice has continued, in large measure, because it is indisputable and roughly correlated with other rankings by the NRC.

The representation of women in academic chemistry is illustrated in Figure 2. While women in 2010 comprised 50.8% of the U.S. population (*18, 21*) and earned nearly the same proportion (49.9%) of chemistry bachelor's degrees (*18*), this is one of the only cases where educational or professional attainment by members of diverse groups approaches parity. Female representation drops significantly at the doctoral level, where 37.8% of chemistry Ph.D.s were earned by women in 2011 (*18*). In the OXIDE chemistry faculty demographics survey for the 2010-11 academic year, we found that women comprised 16%-17% of faculty members in the top 10, top 25, top 50 and top 75 chemistry departments (*22, 23*). These percentages increased to 17-18% in the 2012-13 academic year, based on data from our more recent survey (*24, 25*). The slow growth in representation—a pace that has been roughly consistent for at least a decade (*25*)—is still far from parity, indicating that educational and professional processes for women in academic chemistry are still not equitable.

The prevalence of under-represented racial and ethnic minorities—that is, African-Americans, Hispanics or Latinos/as, and Native Americans—in the U.S. and in academic chemistry is illustrated in Figure 2. The representation with respect to individual and collective groups remains even further from parity than for female scientists. Members of these groups collectively encompassed 29% of the U.S. population in 2010 (*18, 21*), yet they represented just 15% and 11% of the chemistry bachelor's and doctoral degrees earned by U.S. citizens and permanent residents in 2010 and 2011, respectively (*18*). The gap widens even more among tenure-track faculty at top fifty chemistry departments, where under-represented racial and ethnic minorities comprised 5% of assistant and associate professors and 3% of full professors in 2007 (*20*). Figure 2 illustrates comparable trends for the individual racial and ethnic groups. This mismatch in representation becomes even more critical when we consider that the white, non-Hispanic portion of U.S. population is on the decline: It was 69.1% in 2000 (*18, 21*), dropped to 63.7% in 2010 (*18, 21*), and is expected to fall below 50% circa 2043 (*26*). As the U.S. will become a "majority-minority" nation (i.e., no one race or ethnicity exceeding 50% of the population) by mid-century (*26*), it is increasingly urgent that we draw equitably from this ever-growing section of the talent pool. Indeed, the dramatic drop in participation by under-represented racial and ethnic minorities between the doctoral degree and assistant professor ranks suggests that there is a substantial difficulty in bridging the gap through postdoctoral positions; this was strongly highlighted in the report from the Workshop on Achieving Racial and Ethnic Equity in Chemistry (*2*).

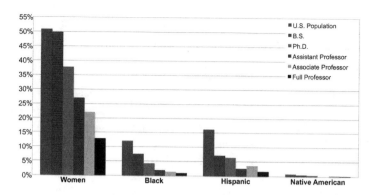

Figure 2. An amalgam of the comparative demographics in the U.S. population (red bars) (18), B.S. chemists (blue bars) (18), Ph.D. chemists (orange bars) (18), Assistant Professors in chemistry (purple bars) (20, 22, 23), Associate Professors in chemistry (green bars) (20, 22, 23), and Full Professors in chemistry (black bars) (20, 22, 23) with respect to women, blacks, Hispanics and Native Americans.

Specific and accurate demographic data regarding disability status is unfortunately difficult to come by because of numerous reasons that include under-reporting due to stigma and disparate and/or evolving definitions and data collection methods (3). Available data regarding the demographics of individuals who report having one or more disabilities include:

- approximately 10% of the civilian, non-institutionalized U.S. population aged 18-64 in 2012 (18);
- 11% of undergraduates in science, technology, engineering, and math (STEM) fields circa 2009 (3);
- 3.4% of chemists earning doctoral degrees in 2011 (18); and
- approximately 8% of STEM Ph.D.s employed as professors in universities and 4-year colleges in 2010 (18).

Paradoxically, the prevalence of disabilities increases at the faculty level; this has been attributed to the fact that disabilities often manifest themselves or are diagnosed as people age (3).

Current estimates suggest that 3-4% of the U.S. population self-identifies as lesbian, gay, bisexual, or transgender (18, 27). Unfortunately, chemistry- or even STEM-specific data with respect to sexual orientation and gender identity seem to be largely unavailable at all educational and professional levels.

Academic Jobs Are Rare Events

An objective of the OXIDE program is the alignment of the demographics of academic chemistry faculties so as to be in proportion with national demographics. One obstacle, however, is that the number of tenure-track/tenured faculty positions is relatively small. For example, in the top fifty chemistry departments, there are approximately 1,600 tenure-track professors (22–25). As such, doubling the percentage in a category with 1% representation would require the addition of only 16 faculty members, which is on the order of a typical year-to-year fluctuation in the total number of faculty. However, the data summarized above indicate that the participation gaps for under-represented groups at nearly all post-secondary educational and faculty levels are currently substantially larger than can be accounted for by these small fluctuations. Consequently, there is a significant need to address the disparity.

The small and finite number of faculty positions presents an additional hurdle. In 2003, there were 6,268 B.S. chemistry degrees awarded, 1,029 Ph.D. chemistry degrees awarded, and roughly 50 new tenure-track positions advertised by the top fifty chemistry departments (2). These are clearly diminishing odds for success along the career path towards a faculty position at one of the leading research-intensive chemistry departments. The good news is that there are a large number of rewarding career paths available to chemistry Ph.D. holders. However, the relative rarity of professorial positions in the leading research-active chemistry departments translates into an additional barrier for any young chemist considering this option as a career track. Anecdotal evidence suggests that this barrier is more severe for members of under-represented groups because they are also being highly recruited by other sectors (2, 15).

Meritocracy and Bias

There is no question that academic departments and their faculty should be driven by rigorous standards of quality and excellence. Publications, citations, grants, patents, awards and other quantifiable metrics can be used to frame the quality of past accomplishments and form the basis for predicting future success. However, even these metrics require interpretation and therein lies the possibility for bias or implicit bias to cloud meritocratic decisions. Social scientists and lawyers have been aware of the role of implicit (or unconscious) bias for some time (1, 2, 28, 29). Much of our social behavior is driven by learned stereotypes that operate automatically—and therefore unconsciously—when we interact with other people. Some of these implicit biases are actually useful in the sense that they represent our past experiences with aggregated empirical metrics and their correlations to future success. The problem occurs when those implicit biases are not connected to objective standards of success. Such implicit biases may in turn lead to decisions about hiring and promotion that are inequitable with respect to diversity (30–39). Implicit bias tests (40) (from Project Implicit, for example) provide a way for individuals to assess the degree to which they hold implicit biases associated with given stereotypes and to form a basis for studying the presence of implicit biases within various cohorts.

Scientists often pride themselves on their ability to think logically and analytically, as they have used such tools successfully in advancing their science. Consequently, they often believe themselves immune to the effects of implicit bias. The implicit bias tests cited above provide a useful tool for most scientists to learn to appreciate the error in this hypothesis. Implicit bias has been shown to play a role in many situations—such as in letters of recommendation (*30*), evaluation of curricula vitae and resumes (*31–35*), evaluation of teaching credibility (*36*), funding of research grant applications (*37*, *38*), and leadership (*39*)—that contribute to the diversity inequities that have historically led to underrepresentation of individuals from certain groups within chemistry faculties. As scientists in an academic setting, we thus need to be particularly vigilant about the impacts of implicit bias and work towards mitigating them within all of our interactions with students and colleagues of any background.

Other Barriers to Diversity Equity

Beyond implicit bias, many other factors can also serve as inequitable barriers to the success of each member of a diverse work group. Handelsman and coworkers (*41*), for example, wrote an influential article listing the significant negative impact of bias and campus climate on the careers of female faculty. Unfortunately, individuals from diverse backgrounds face these and many more barriers due to myriad factors that have been identified in the social science literature:

- schemas and stereotype threat
- accumulation of bias
- lack of universal design
- insufficient mentoring
- insufficient/unequal "family-friendly" policies
- overburdening the under-represented
- unwelcoming/non-accommodating climate
- unwelcoming/non-accommodating professional cultures
- qualitative vs. quantitative assessment
- solo status
- minimizing differences/colorblindness
- depoliticization and meritocratic ideology

Unfortunately, our space here does not allow us to discuss these topics in depth. Interested readers are encouraged to use these as search terms in their further study.

The role that some of these issues play in creating unaccommodating environments for individuals from diverse backgrounds are often not intentional, but they are nevertheless harmful. Any one of these barriers would likely be insignificant on its own, but their accumulation tends to deter diverse participation (*29*).

One such example lies in the possible marketing strategy by a department to feature pictures of prominent individuals from the past—whether they be chemists from their department or more broadly—on a website or on the walls of their building. Such advertising is well-intentioned, as it seeks to show the accomplishments of chemists and the possibilities that a chemistry career would afford a young person. However, at the present moment, the bulk of past scientists are white, male, seemingly heterosexual and apparently able-bodied. As such, this strategy also promotes the unintended message that chemistry is not a career choice for individuals from diverse groups.

Alternatively, a department may choose to publicly position itself as following "colorblind" practices, meaning that it explicitly and consciously claims not to see race and ethnicity when making meritocratic decisions. However, these policies can generate ironic effects for individuals who may feel that their sense of identity is being stripped away and that the diversity they add to the organization is not being valued. Thus, "colorblind" policies have been seen to be counterproductive in the sense of repelling diverse participation (*42*, *43*).

Yet another possibly ironic action undertaken by departments attempting to improve the diversity climate is requiring participation in diversity training, perhaps in tandem with the annual lab safety training that is common practice in chemistry. Unfortunately, the success of diversity training as typically implemented in the business sector has been seen to range from ineffective to even counterproductive with respect to increasing diverse participation (*44*). This result is perhaps not surprising to chemists because safety training, when done poorly, also has a tendency to have little impact on changing the safety culture in departments. Consequently, it is not enough to simply institute an external diversity training program. Instead, departments must embrace an internal program that is effective and inclusive of all their members.

In summary, the barriers to diversity equity are unfortunately large in number. While these barriers can originate from issues existing prior to their identification, they can also arise as unintended consequences of some attempts to redress inequities.

Removing Barriers

In order to remove the barriers to equitable participation, it is important to first identify them. This therefore drives a need to discuss issues such as implicit bias and the damaging effects it can have in unfairly assessing individuals from different backgrounds. It is also useful to raise awareness of and address the many possible barriers encountered by individuals from particular under-represented groups, as discussed above. For example, the Workshop on Gender Equity (*1*) identified the lack of campus daycare facilities—and family-friendly practices in general—as a significant barrier to the success of tenure-track female chemists. While this issue was particularly relevant in the area of gender equity, the solutions enacted—for example, the introduction of day care services on many campuses or individualized arrangements subsidized by institutional funds—have helped to create a more accommodating environment for all faculty. Another example lies in the tenure decision, which is a barrier that all tenure-track faculty face. Historically, the standards for success have been given in relative and opaque terms. The problem is that the lack of clear dialogue and transparency in discussing these standards does not promote inclusive excellence (1, 2). (It is notable that this observation appears to be true whether we are speaking about the metrics to assess promotions along the tenure track or the quality of a department as a whole.) As a consequence, many chemistry departments and universities are now providing clear outlines for the tenure and promotion process and statements about the standards that will be used in making such decisions.

One possible approach toward identifying and correcting remaining diversity inequities would hinge on the passive offer to remove barriers as they are identified by individuals within a department. Such an approach would perhaps be well-intentioned and motivated by the argument that there is no way to know what particular individuals will need. However, it is unlikely that individuals near the bottom of the academic ladder —such as graduate students or assistant professors—would risk their tenuous positions by making such requests. Instead, department administrators and thought leaders need to take a proactive approach, learning about and changing policies and practices that have already been identified as creating diversity inequities. While this top-down approach can lead to much better departmental climates, it requires active engagement by chemistry department chairs or heads.

The Open Chemistry Collaborative in Diversity Equity

OXIDE was formed precisely to change the academic chemistry infrastructure from the top down so as to flatten diversity inequities in academic chemistry departments. We partner with department chairs, placing both the responsibility and the credit for solving the problem on them, rather than on single change agents in the rank and file. Our top-down approach complements existing and successful bottom-up programs such as those by SACNAS (45), AISES (46), NOBCChE (47), NOGLSTP (48), COACh (49), and others. We are helping to create an accommodating and inviting climate in leading research-intensive chemistry departments such that individuals from diverse backgrounds mentored and trained through the bottom-up approach will choose to remain as tenure-track faculty in leading departments.

The key elements of the OXIDE strategy are summarized as follows:

1. Change academic chemistry infrastructure from the top down.
2. Be excellence-driven; diversity is key to the post-modern meritocracy.
3. Partner with chairs of research-intensive chemistry departments.
4. Assign responsibility and give credit to the partners.
5. Focus on reducing inequitable policies and practices that have historically led to disproportionate representation.
6. Collaborate with social scientists, who have a broader knowledge of diversity and inclusion.
7. Focus on diversity writ large (e.g., gender and gender identity, race-ethnicity, disabilities, and sexual orientation).

8. Disseminate information about inclusive excellence and diversity broadly.

These elements are infused in OXIDE's two most visible activities: the annual faculty demographics surveys and the biennial NDEWs. In order to maintain dialogue with chemistry faculties and their chairs between NDEWs, OXIDE relays news by email, updates material and information on our website (http://www.oxide.gatech.edu), delivers presentations in departments and at (inter)national meetings, and engages in conversation with chairs and their representatives though virtual and physical meetings.

OXIDE conducts faculty demographics surveys to collect data on the year-to-year demographics of tenure-track/tenured faculty at leading research-intensive chemistry departments, which are defined according to the NSF rankings of chemical research expenditures mentioned earlier in this chapter. Department chairs or their designees report the aggregate demographics, by rank, of the faculty who meet the above criteria and who are primarily (i.e., greater than 50%) appointed in the department. These criteria have been chosen for consistency with earlier data sets (20). Gender data for the academic years 2009-10 (22, 23), 2010-11 (22, 23), 2011-12 (24, 25) and 2012-13 (24, 25) have been gathered and reported in *Chemical & Engineering News* and on the OXIDE website. (We use the term 'gender' rather than 'gender identity' here because we received no data about individuals who define their gender in ways besides the male-female binary, although we afforded chairs an opportunity to provide such data in the 2011-13 survey.) Race-ethnicity data were also collected for academic years 2011-12 and 2012-13, and we plan to disseminate this data in 2014. This longitudinal repository of demographic data, which we plan to continue building annually, allows the chemistry community's progress to be measured as we (hopefully) shrink the gap in representation between chemistry faculty and the general public. In the future, we plan to carefully explore possible methods for gathering data about faculty sexual orientation and disability status while keeping in mind the particular sensitivities and privacy concerns associated with data collection related to these demographic categories.

The most recent NDEW was held April 15-16, 2013 in Arlington, VA. More than 70 participants—including chairs and designated representatives from over 35 chemistry departments, social scientists, and representatives from federal agencies, foundations, and diversity communities—listened attentively to lectures from social scientists and others, engaged in breakout sessions, and collaborated on possible solutions to address diversity inequities. For the first time, an entire session was devoted to discussion of the diversity inequities faced by LGBTQIQ chemists and the actions that can be taken by departments to create a more inclusive climate in this respect. A session on organizational structure, behavior and dynamics discussed research-based strategies for mitigating implicit bias, stereotype threat and colorblindness. Effective practices were provided in a session on recruitment, hiring, retention, and promotion. The context of professional cultures and how they can reinforce inequalities played a critical role in a session on creating an inclusive climate. In addition to the primary sessions, NDEW2013 engaged participants in a series of breakout groups. Each group was assigned a different topic, article or website to motivate its proposal for new practices and policies that could be implemented to address an existing shortfall in diversity equity. Post-workshop assessment data indicate that numerous departments have begun to implement programs and practices to reduce diversity inequities as a result of their participation.

One outcome of NDEW2013 was the construction of a set of recommendations intended for immediate implementation by the chairs:

1. Conduct a faculty meeting on diversity excellence:

    a. Walk your faculty through the generic department presentation given at NDEW2013 (or a version customized to your department).
    b. Make sure that you do not advertise the event as diversity training.
    c. Emphasize strategies that mitigate stereotype threat.

2. Create mentoring programs (vertical and horizontal).
3. Create a department diversity committee.

    a. It should be broadly reflective of your faculty's perspectives (for example, include straight, able-bodied white male faculty in addition to faculty from under-represented groups).
    b. Don't overburden faculty from under-represented groups.
    c. Establish deliverables to measure the committee's success.

4. Conduct faculty searches in broad areas.
5. Respond to current and future OXIDE surveys (for example, on workshop evaluation, demographics, and climate).
6. Implement a policy/program targeted to address climate and/or demographics. If possible, partner with OXIDE to assess it.

These are a minimal set of actions that departments can make to begin changing the climate towards inclusive excellence. We are happy to report that several of the participating departments have adopted all (or nearly all) of these suggestions (though precise statistics are pending, as analysis of the NDEW follow-up data is in progress).

Many of these suggestions should require little explanation to the reader, based on what has already been discussed in this chapter. However, the fourth suggestion (*50*) may seem surprising. It is based on the fact that the numbers of individuals in faculty applicant pools are integers and that the numbers of under-represented individuals in such pools tend to be particularly small integers. Consequently, if the area of expertise for a particular faculty search is defined too narrowly, there is an increased possibility that the number of qualified applicants from under-represented groups will be very small and perhaps equal to zero. Meanwhile, an outstanding candidate in a different research field who would be overlooked by this narrow search strategy might not remain in the job pool for a subsequent year's search and may even leave academia entirely. Departments should therefore not limit themselves to conducting annual searches based solely on area of research expertise. Instead, they should focus on the best candidates available in any given year and think more broadly in terms of how these candidates fit the needs of the department's overall research portfolio across several search years. In so doing, departments are more likely to enhance both their diversity and quality.

Readers interested in more details about OXIDE and our activities can refer to our website, http://oxide.gatech.edu.

Conclusions

In summary, by working with the chairs of leading research-intensive chemistry departments, OXIDE aims to reduce inequitable policies and practices that have historically led to disproportionate representation on academic faculties with respect to gender and gender identity, race-ethnicity, disabilities, and sexual orientation. OXIDE and its partners are driven by the notion that the meritocratic underpinning of research-intensive chemistry departments will be stronger if conjoined with the notion of inclusive excellence. Indeed, all the elements that make a chemistry department more diverse are critical to advancing its mission and increasing its academic prestige. The chairs and thought leaders within chemistry departments play a critical role in creating the climate from the top down. As such, OXIDE's role is that of a connector between department leaders, diversity communities and the social sciences so as to identify and remove barriers to diversity equity.

Acknowledgments

This work and the OXIDE program have been jointly supported by the NIH, DOE and NSF through NSF grant #CHE-1048939. Cognizant units are the Pharmacology, Physiology, and Biological Chemistry Division at the National Institute of General Medical Sciences (NIGMS) of the National Institutes of Health (NIH), the Office of Basic Energy Sciences (BES) at the Department of Energy (DOE), and the Chemistry Division of the Math and Physical Sciences Directorate (MPS) at the National Science Foundation (NSF). The Georgia Tech Center for the Study of Women, Science, and Technology (WST) has supported OXIDE's undergraduate assistants through their Student-Faculty Research Partnerships program. It is a pleasure to thank the many colleagues, collaborators, and board members who have helped to advance the OXIDE mission since its inception as the DEWPC. Among these are Christopher Bannochie, Rudy Baum, Caroline Bertozzi, Karl Booksh, Sheila Browne, Larry Dalton, Frank Dobbin, Luis Echegoyen, Archie Ervin, Joseph Francisco, Michelle Francl, Aliola Gardner-Aben, Paula Hammond, Kendall Houk, Lisa Hwang, Malika Jeffries-EL, Sharon Neal, Mary Jo Ondrechen, Jonathan Pang, Geraldine Richmond, Maureen Rouhi, Sophie Rovner, Denise Sekaquaptewa, Erik Sorensen, Jean Stockard,

Timothy Swager, Isiah Warner and Alveda Williams. We are also delighted to thank H. N. Cheng and Marinda Wu for encouraging us to write this chapter.

References

1. Friend, C. M.; Houk, K. N. Workshop on Building Strong Academic Chemistry Departments through Gender Equity, Arlington, VA, Jan. 29–31, 2006.
2. Warner, I. M.; Turro, N. J. Workshop on Excellence Empowered by a Diverse Workforce: Achieving Racial and Ethnic Equity in Chemistry, Arlington, VA, Sept. 24–26, 2007.
3. Bowman-James, K.; Benson, D.; Mallouk, T. Workshop on Excellence Empowered by a Diverse Academic Workforce: Chemists, Chemical Engineers & Materials Scientists with Disabilities, Arlington, VA, Feb. 8–10, 2009.
4. *A Data-Based Assessment of Research-Doctorate Programs in the United States*; Ostriker, J. P., Holland, P. W., Kuh, C. V., Voytuk, J. A., Eds.; The National Academies Press: Washington, DC, 2011.
5. Patton, S. Letter to the Editor: Advice for the Young Women of Princeton: The Daughters I Never Had. *Daily Princetonian*; March 29, 2013.
6. Slaughter, A.-M. Why Women Still Can't Have It All. *The Atlantic*; June 13, 2012.
7. Sandberg, S. *Lean in: Women, Work, and the Will to Lead*; Alfred A. Knopf: New York, NY, 2013.
8. Hoobler, J. M.; Lemmon, G.; Wayne, S. J. Women's Underrepresentation in Upper Management: New Insights on a Persistent Problem. *Organizational Dynamics* **2011**, *40*, 151–156.
9. Gilgoff, D. Breaking the Corporate Glass Ceilings. *U.S. News & World Report*; Nov. 20, 2009.
10. Davies, C. PM's Apology to Codebreaker Alan Turing: We Were Inhumane. *The Guardian*; Sept. 10, 2009.
11. Robinson, M.; Pfeffer, C.; Buccigrossi, J. *Business Case for Inclusion and Engagement 2003*; http://workforcediversitynetwork.com/docs/business_case_3.pdf (accessed June 24, 2014.)
12. Hubbard, E. E. *The Diversity Scorecard: Evaluating the Impact of Diversity on Organizational Performance (Improving Human Performance)*; Elsevier: Oxford, U.K., 2004.
13. Jackson, S. E. *Diversity in the Workplace: Human Resources Initiatives*; Guilford Press: New York, 1992.
14. Page, S. E. *The Difference: How the Power of Diversity Creates Better Groups, Firms, Schools, and Societies*; Princeton University Press: Princeton, NJ, 2007.
15. Francisco, J. S.; Warner, I. M. *Minorities in the Chemical Workforce: Diversity Models that Work - A Workshop Report to the Chemical Sciences Roundtable*; The National Academies Press: Washington, DC, 2003.

16. Watkins, S. F. The Imperative of Leaders and Organizations. In *Minorities in the Chemical Workforce: Diversity Models That Work*; The National Academies Press: Washington, DC, 2003; pp 66−79.
17. Laursen, S. L.; Weston, T. J. Trends in Ph.D. Productivity and Diversity in Top-50 U.S. Chemistry Departments: An Institutional Analysis. *J. Chem. Educ.* **2014** DOI: 10.1021/ed4006997. Published Online: June 3, 2014 (accessed June 24, 2014).
18. *Women, Minorities, and Persons with Disabilities in Science and Engineering: 2013*, Special Report NSF 13-304; National Center for Science and Engineering Statistics, National Science Foundation: Arlington, VA, 2013. http://www.nsf.gov/statistics/wmpd/. (Data collected for 2010 to 2012, with some data updated in May 2014.)
19. Nelson, D. J. Diversity in Academia: A Look at Gender and Race/Ethnicity in Science and Engineering Departments. *FASEB J.* **2002**, *16A* (4), 524.
20. Nelson, D. J. The Nelson Diversity Surveys; Norman, OK. http://faculty-staff.ou.edu/N/Donna.J.Nelson-1/diversity/Faculty_Tables_FY07/ChemTable2007.pdf.
21. U.S. Census Data. http://www.census.gov.
22. Rovner, S. Women Are 17% Of Chemistry Faculty. *Chem. Eng. News* **2011**, *89* (44), 42–46.
23. Hernandez, R. *Faculty Gender Demographics of the Leading Research-Active Chemistry Departments: Academic Years (AY) 2009-10 and 2010-11*, OXIDE. http://www.oxide.gatech.edu/ver1.0/data/demographics.html.
24. Watt, S.; Hernandez, R. *Faculty Gender Demographics of the Leading Research-Active Chemistry Departments: Academic Years (AY) 2011-12 and 2012-13*, OXIDE. http://www.oxide.gatech.edu/ver1.0/data/demographics.html.
25. Rovner, S. Women Faculty Positions Edge Up. *Chem. Eng. News* **2014**, *92* (14), 41–44.
26. *U.S. Census Bureau Projections Show a Slower Growing, Older, More Diverse Nation a Half Century from Now*; U.S. Census Bureau: Washington, DC, December 12, 2012. https://www.census.gov/newsroom/releases/archives/population/cb12-243.html.
27. Gates, G. J.; Newport, F. Special Report: 3.4% of U.S. Adults Identify as LGBT. Gallup: 2012, Gallup: Inaugural Gallup findings based on more than 120,000 interviews. http://www.gallup.com/poll/158066/special-report-adults-identify-lgbt.aspx.
28. Greenwald, A. G.; Krieger, L. H. Implicit Bias: Scientific Foundations. *California Law Review* **2006**, *94* (4), 945–967.
29. Valian, V. *Why So Slow? The Advancement of Women.*; MIT Press: Cambridge, MA, 1998.
30. Trix, F.; Psenka, C. Exploring the Color of Glass: Letters of Recommendation for Female and Male Medical Faculty. *Discourse & Society* **2003**, *14* (2), 191–220.
31. Bertrand, M.; Mullainathan, S. Are Emily and Greg More Employable than Lakisha and Jamal? A Field Experiment on Labor Market Discrimination. *American Economic Review* **2003**, *94* (1), 991–1013.

32. Moss-Racusin, C. A.; Dovidio, J. F.; Brescoll, V. L.; Graham, M. J.; Handelsman, J. Science faculty's subtle gender biases favor male students. *Proc. Natl. Acad. Sci. U.S.A.* **2012**, *109* (41), 16474–16479.
33. Steinpreis, R. E.; Anders, K. A.; Ritzke, D. The Impact of Gender on the Review of the Curricula Vitae of Job Applicants and Tenure Candidates: A National Empirical Study. *Sex Roles* **1999**, *41* (7–8), 509–528.
34. Tilcsik, A. Pride and Prejudice: Employment Discrimination against Openly Gay Men in the United States. *Am. J. Sociol.* **2011**, *117* (2), 586–626.
35. Correll, S. J.; Benard, S.; Paik, I. Getting a Job: Is There a Motherhood Penalty? *Am. J Sociol.* **2007**, *112* (5), 1297–1338.
36. Russ, T. L.; Simonds, C. J.; Hunt, S. K. Coming Out in the Classroom... An Occupational Hazard?: The Influence of Sexual Orientation on Teacher Credibility and Perceived Student Learning. *Commun. Educ.* **2002**, *51* (3), 311–324.
37. Ginther, D. K.; Schaffer, W. T.; Schnell, J.; Masimore, B.; Liu, F.; Haak, L. L.; Kington, R. Race, Ethnicity, and NIH Research Awards. *Science* **2011**, *333*, 1015–1019.
38. Wenneras, C.; Wold, A. Nepotism and sexism in peer-review. *Nature* **1997**, *387*, 341–343.
39. Porter, N.; Geis, F. Women and Nonverbal Leadership Cues: When Seeing Is Not Believing. In *Gender and Nonverbal Behavior*; Mayo, C., Henley, N. M., Eds.; Springer Verlag: New York, 1981; pp 39–61.
40. Greenwald, A. G.; Nosek, B. A.; Banaji, M. R. Understanding and using the Implicit Association Test: I. An improved scoring algorithm. *J. Personality Social Psychol.* **2003**, *85*, 197–216.
41. Handelsman, J.; Cantor, N.; Carnes, M.; Denton, D.; Fine, E.; Grosz, B.; Hinshaw, V.; Marrett, C.; Rosser, S.; Shalala, D.; Sheridan, J. More Women in Science. *Science* **2005**, *309* (5378), 1190–1191.
42. Purdie-Vaughns, V.; Steele, C. M.; Davies, P. G.; Ditlmann, R.; Crosby, J. R. Social Identity Contingencies: How Diversity Cues Signal Threat or Safety for African Americans in Mainstream Institutions. *J. Personality Social Psychol.* **2008**, *94*, 615–630.
43. Plaut, V. C.; Thomas, K. M.; Goren, M. J. Is Multiculturalism or Color Blindness Better for Minorities? *Psychol. Sci.* **2009**, *20*, 444–446.
44. Kalev, A.; Dobbin, F.; Kelly, E. Best practices or best guesses? Assessing the efficacy of corporate affirmative action and diversity policies. *Am. Soc. Rev.* **2006**, *71*, 589–617.
45. SACNAS: Society for Advancement of Chicanos and Native Americans in Science. http://sacnas.org.
46. AISES: American Indian Science and Engineering Society. http://www.aises.org.
47. NOBCChE: National Organization for the Professional Advancement of Black Chemists and Chemical Engineers. https://www.nobcche.org.
48. NOGLSTP: National Organization of Gay and Lesbian Scientists and Technical Professionals. http://www.noglstp.org.
49. COACh: Committee on the Advancement of Women Chemists. http://coach.uoregon.edu/.

50. University of Michigan Gender in Science and Engineering, Report of the Subcommittee on Faculty Recruitment, Retentions and Leadership. http://www.advance.rackham.umich.edu/GSE-_Faculty_Recruitment_Retention.pdf.

Chapter 20

Diversity and Departmental Rankings in Chemistry

Cedric Herring*

Department of Sociology (M/C 312), University of Illinois at Chicago, 1007 West Harrison, Chicago, Illinois 60607, and Institute of Government and Public Affairs (M/C 191), University of Illinois, 815 West Van Buren, Suite 525, Chicago, Illinois 60607
*E-mail: herring@uic.edu; telephone: 312/304-1515

This work examines whether racial and gender diversity are related to the rankings of chemistry and chemical engineering programs at research universities. It uses data from the 2011 National Academy of Sciences (NAS) Rankings of U.S. Research Universities to examine the competing expectations of the "value-in-diversity perspective" and the "diversity as process-loss" perspective. The results are counter to the expectations of the diversity as process-loss perspective, which predicts that diversity (and efforts to achieve it) are *harmful* to organizations. The results are fully consistent with the value-in-diversity perspective, which predicts that as organizations become more diverse, they benefit relative to their competitors. The results show that, net of faculty publication rates, visibility of faculty publications, percentage of faculty with grants, percentage assistant professors, program faculty size, region, whether the institution is public or private, number of student activities, financial support for students, and number of PhD graduates, diversity is positively associated with departmental rankings of chemistry and chemical engineering programs in research universities. The implications of these findings for diversity in chemistry are discussed.

Keywords: diversity; departmental rankings; research productivity; race; gender; research grants

Introduction

As the demography of the nation continues to change, diversity has become a critically important topic that poses philosophical, political and policy challenges. In business and industry, it has become common for proponents of the "business case for diversity" to claim that "diversity pays (*1–3*)." Proponents of this view suggest that diversity in the corporate setting, in particular, represents a compelling business interest that will help meet customers' needs, enrich understanding of the pulse of the marketplace, and improve the quality of products and services offered (*4*).

However, the rhetoric about diversity in higher education is different (*5*). Rather than putting forth the business case, proponents of inclusion have argued that it is necessary to safeguard diversity in America's institutions of higher education for the sake of their students. Typically, university officials and others have argued for affirmative action because such policies allows students "to live and study with classmates from a diverse range of backgrounds, [which] is essential to students' training for this new world, nurturing in them an instinct to reach out instead of clinging to the comforts of what seems natural or familiar" (*6*). In other words, the central argument has been that universities have a compelling interest in promoting diversity because their students benefit from their doing so.

The purpose here is not to argue that students do not benefit from diversity. They do (*7*). But universities may have a compelling interest in promoting diversity for reasons very similar to businesses. In particular, it is possible that research universities benefit directly from diversity—much like business organizations—because diversity (among their faculty and their students) enhances their reputational bottom lines. As Lee Bollinger puts it, "[u]niversities understand that to remain competitive, their most important obligation is to determine — and then deliver — what future graduates will need to know about their world and how to gain that knowledge. While the last century witnessed a new demand for specialized research, prizing the expert's vertical mastery of a single field, the emerging global reality calls for new specialists who can synthesize a diversity of fields and draw quick connections among them" (*6*).

Using data from the 2011 National Academy of Sciences (NAS) Rankings of U.S. Research Universities, this work examines whether racial and gender diversity "pay" in terms of the rankings of chemistry and chemical engineering programs at research universities. The NAS dataset consists of several indicators relating to research productivity, student support and outcomes, and program diversity from 278 chemistry and chemical engineering doctoral programs at U.S. research universities. This work examines the relationship between departmental rankings at research universities and diversity, net of factors such as program size, region, publication rates, grants, scholarly awards, and whether the institution is public or private.

Diversity in Chemistry

Figure 1 shows that according to data from the National Science Foundation and the National Academy of Science, in 2008 50% of those receiving bachelor's degrees in chemistry were female. However, when looking at the chemistry pipeline, it is clear that women become increasingly underrepresented. In particular, 41.6% of graduate students in chemistry are female; 36.1% of those receiving doctorates in chemistry are female; but only 13.8% of those on faculty at research universities are female.

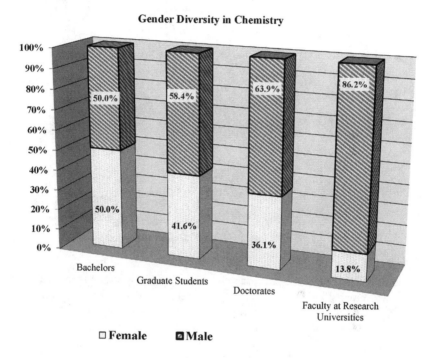

Figure 1

Similarly, Figure 2 shows that in 2008 21.4% of those receiving bachelor's degrees in chemistry were underrepresented minorities. However, when one looks at the chemistry pipeline, it is clear that minorities become increasingly underrepresented. In particular, 19.8% of graduate students in chemistry are underrepresented minorities; 19.2% of those receiving doctorates in chemistry are underrepresented minorities; but only 4.6% of those on faculty at research universities are underrepresented minorities.

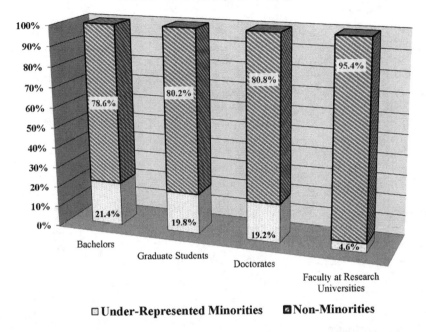

Figure 2

Departmental Rankings in Chemistry: Does Diversity Pay?

Academic departments in research universities strive for preeminence. In chemistry, much like in other disciplines, there has been a growing obsession with the rankings of university programs (8, 9). Most scholars are critical of how such rankings are derived (10); nevertheless, they are aware of the impact and consequences—both intended and unintended—of such rankings (10, 11). As intended, these rankings help prospective students choose where to invest their educational dollars to attend college (10). According to Espeland and Sauder, "[m]ost students believe the reputation of the school is an important determinant of career trajectories" (10). Such rankings also help universities identify their areas of strength and weakness (10).

Espeland and Sauder also point out some additional consequences of such rankings (10):

> Rankings are designed to be reflections of existing school characteristics and quality, to report—in a disinterested and objective fashion—how schools compare to each other on selected criteria. However, we've found that rankings actually shape the hierarchy of the institutions they're trying to assess. Over time, schools change their activities and policies to optimize their standings on the [ranking] criteria.... Rankings influence who is admitted to which schools, how scholarship money is allocated, which programs are well-funded and which aren't, as well as other serious

forms of redistribution of both resources and opportunities. Rankings are also used to fire and reward administrators, allocate budgets across universities, and may even challenge the mission of schools whose goals aren't captured in rankings factors (10).

The National Academy of Sciences, in perhaps the most respected recent study of departmental rankings, found that research rankings, publications per allocated faculty member, citations per publication, and grants per allocated faculty were highly related to reputational rankings in research departments (12). Baldi uncovered the advantages of size and age (13).

Diversity, however, is an often overlooked factor. For many people, diversity is a good thing because it helps remove barriers that have historically prevented access to people of color, women, members of the lesbian, gay, bisexual, and transgender (LGBT) community, the disabled, and others who have been under-represented in the corridors of power.

Top-ranked programs are concerned with how they can maintain their rankings. Lower ranked programs want to know how they can climb in the rankings. This raises the question of what effect, if any, does diversity have on the rankings of chemistry departments? Does diversity help, hurt, or have no effect on departmental rankings in chemistry?

Using data from the 2011 National Academy of Sciences (NAS) Rankings of U.S. Research Universities, this work examines the effect of racial and gender diversity on the rankings of chemistry and chemical engineering programs at research universities, net of other factors that have been associated with reputational rankings. The NAS dataset offers the opportunity to examine these factors in a transparent way, as it consists of several indicators relating to research productivity, student support and outcomes, and program diversity from 278 chemistry and chemical engineering doctoral programs at U.S. research universities. It allows for the examination of the relationship between reputational rankings of chemistry and chemical engineering departments at research universities and diversity, net of factors such as publication rates, visibility of faculty publications, grants, program faculty size, region, and whether the institution is public or private. Below, this work provides two competing views about the likely impact of diversity on reputational rankings in chemistry: (1) the value-in-diversity perspective and (2) the diversity as process-loss perspective.

Proponents of the Value-in-Diversity Perspective

In business, diversity produces better business results (14). Exploiting the nation's diversity is viewed as key to future prosperity. Ignoring the fact that discrimination limits a society's potential because it leads to underutilization of talent pools is no longer practical nor feasible. Diversity campaigns, thus, are part of the attempt to strengthen the United States and move beyond its history of discrimination by providing previously excluded groups greater access to institutions and workplaces (15).

But the value-in-diversity perspective is not only applicable to business organizations. As Iverson points out, universities' "diversity action plans describe higher education as a 'highly competitive market'; fierce competition exists in the recruitment of diverse individuals, and institutions strategize about how to maintain a competitive edge in response to 'rapidly changing market conditions' and 'a new demographic reality' in an increasingly global marketplace" (16). Diversity is viewed as being an essential ingredient for realizing and keeping a competitive advantage over other educational institutions.

Prior research has suggested that there is support for the critical diversity and value-in-diversity perspective for business organizations (14). But does such reasoning hold true for organizations that are not primarily profit-oriented such as chemistry programs at research universities? This question is addressed below. First, however, this work discusses the views of skeptics of the value-in-diversity perspective.

Skeptics of the Value-in-Diversity Perspective

Skeptics of diversity are more suspicious of the benefits that diversity provides (17–21). Some skeptics argue that diversity leads to "process loss"–inefficiencies in group process that occur when group members are involved in decision-making (22–24). For example, Skerry points out that research on intergroup relations consistently finds that racial and ethnic diversity are linked with conflict, especially emotional conflict among co-workers (17). Jehn, Northcraft, and Neale found that while informational diversity positively influenced group performance, value and social category diversity diminished this effect (24). Pelled, Eisenhardt, and Xin found a complex link between work group diversity and work group functioning (22). Tsui, Egan, and O'Reilly found that diversity can reduce the cohesiveness of the group and result in increased employee absenteeism and turnover (20). Moreover, detractors of the value-in-diversity model suggest that the emphasis on diversity divides America into separate groups based on race, ethnicity, or gender and in so doing suggests that some social categories are more deserving of privileges than are others (25). There is also the argument that greater diversity is associated with lower quality because it places lower performing people in positions for which they are not suited (18). Finally, Williams and O'Reilly suggest that most empirical evidence suggests that diversity is most likely to hinder group functioning (26). In short, skeptics of diversity suggest that group differences result in conflict and process loss (27).

Many of the claims about the impact of diversity on universities in general, and chemistry departments in particular, have gone unexamined empirically; therefore, it is not clear whether diversity "pays" for such organizations. It is possible that diversity has many benefits as the value-in-diversity perspective suggests. It is also plausible that any benefits of diversity are more than offset by significant costs that have been identified by diversity as a process loss. The growing heterogeneity within chemistry warrants a serious examination of the impact of diversity. Using data from the 2011 National Academy of Sciences (NAS) Rankings of U.S. Research Universities, this analysis examines whether racial and gender diversity among the faculty and graduate students "pay" in terms of the rankings of chemistry and chemical engineering departments at research universities. First, however, the data and analysis strategies are discussed.

Data and Methods

The data for our analysis come from the 2011 National Academy of Sciences (NAS) Rankings of U.S. Research Universities. The NAS Rankings of U.S. Research Universities

provides an unparalleled dataset collected from doctoral institutions, doctoral programs, doctoral faculty and public sources that can be used to assess the quality and effectiveness of doctoral programs based on measures important to faculty, students, administrators, funders, and other stakeholders. The committee collected 20 measures that include characteristics of the faculty, such as their publications, citations, grants, and diversity; characteristics of the students, such as their GRE scores, financial support, publications, and diversity; and characteristics of the program, such as number of PhDs granted over five years, time to degree, percentage of student completion, and placement of students after graduation (National Academy of Sciences) (*12*).

The unit of analysis for the study is the program or department. The analysis uses the NAS data to assess the relationship between program diversity and program ranking. It examines this relationship net of other factors that may be related to program rankings such as number of publications per faculty, citations per publication, percentage of faculty with grants, number of scholarly awards per faculty, number of graduate students, number of full-time equivalent faculty, size of program, region, and whether the institution is public or private. Below, this work describes how the National Academy of Sciences measured these variables.

<u>Program Rankings</u>. For each discipline, the NAS surveyed faculty to obtain their views on different characteristics of programs as measures of quality. Over 87,000 faculty involved in doctoral education were surveyed. Rankings of program quality are reputational; i.e., raters were asked to rate up to 15 programs for "scholarly quality." Based on these ratings, chemistry departments were ranked from 1 through 172, and chemical engineering departments were ranked from 1 through 106. These rankings were converted to percentile ranks; thus, rankings range from 0 (lowest ranking) to 100 (highest ranking).

<u>Diversity</u>. For each department, the race/ethnicity of faculty in the program was used to compute the ratio of non-Hispanic Blacks, Hispanic, and American Indians or Alaska Natives to that of all faculty with known race/ethnicity. In addition, the gender of faculty in the program was used to compute the ratio of female to total faculty in the program. Also, the race/ethnicity of graduate students in the program was used to compute the ratio of non-Hispanic Blacks, Hispanics, and American Indians or Alaska Natives to that of all graduate students with known race/ethnicity. The gender of graduate students in the department was used also to compute the ratio of female to total graduate students in the program. Finally, the percentage of international students was computed by taking the number with temporary visas and dividing it by the number of graduate students with known citizenship status. For the diversity index, departments were then ranked on the basis of their total diversity on all of these dimensions. These rankings were

converted to percentile ranks; thus, rankings range from 0 (lowest ranking) to 100 (highest ranking).

Publications. Number of (verified) published books and articles per allocated faculty, 2001-2006.

Citations. Average (verified) citations per publication.

Grants. Percentage of faculty with grants.

Private. Whether the institution's form of control is public or private (coded 1 for private).

Region. Dummy variable coded to indicate North East (Maine, New Hampshire, Vermont, Massachusetts, Rhode Island, Connecticut, New York, Pennsylvania, and New Jersey); Midwest (Wisconsin, Michigan, Illinois, Indiana, Ohio, Missouri, North Dakota, South Dakota, Nebraska, Kansas, Minnesota, and Iowa); South Atlantic (Delaware, Maryland, District of Columbia, Virginia, West Virginia, North Carolina, South Carolina, Georgia, and Florida); South Central (Kentucky, Tennessee, Mississippi, Alabama, Oklahoma, Texas, Arkansas, and Louisiana); and West (Idaho, Montana, Wyoming, Nevada, Utah, Colorado, Arizona, New Mexico, Alaska, Washington, Oregon, California, and Hawaii).

Assistant Professors. Assistant professors as a percentage of total faculty.

FTE. Number of allocated faculty in 2006 corrected for association with multiple programs.

Number of Student Activities. How many of the 18 different kinds of support activities for doctoral students or doctoral education.

Year 1 Support. Percentage of first year students with full financial support.

PhD Graduates. Average number PhD students graduated per year.

These factors are used in the multivariate analysis that examines the net relationship between diversity and program rankings in chemistry and chemical engineering in research universities.

Results

Several factors contribute to the reputations of chemistry programs in research universities. A department's reputation may be a reflection of its recent research accomplishments. Alternatively, such reputations may be unduly influenced by renowned professors who retired long ago or the major contributions of a few faculty with distinguished records. Still, it is likely that reliable data can help shed light on the significant dimensions of departmental reputations. It is also possible that factors, such as diversity, are often overlooked.

How is diversity related to the rankings of chemistry and chemical engineering departments in research universities? Table 1 presents selected characteristics for low-ranking, medium-ranking, and high-ranking programs. This table shows that on average, those departments with high program rankings have mean number of publications per faculty that are more than twice those of low-ranking programs. The citation rates of low-ranking programs are 55% of those of high-ranking programs. About 69% of faculty in low-ranking programs have grants compared with 89% of faculty in high-ranking programs. High-ranking programs are about twice as likely to be in private universities compared with those with low rankings. High-ranking programs are over-represented in the Northeast but under-represented in the South Central. Low-ranking programs are over-represented in the South Central. On average, high-ranking programs have a slightly lower percentage of assistant professors, have slightly more faculty, offer more support activities to their graduate students, offer support to a higher proportion of their first-year students, and graduate more students per year. Generally, low-ranking programs are at the opposite end of the spectrum. Such differences suggest that these factors should be statistically controlled when examining the relationship between diversity and program rankings.

Table 1. Selected Characteristics of Chemistry and Chemical Engineering Programs by Ranking

Selected Characteristics	Department Ranking			
	Low	Medium	High	Total
Mean Number of Publications	1.4	2.2	3.7	2.4
Citations per Publication	1.5	2.2	2.7	2.2
% with Grants	69.5%	80.7%	89.3%	79.9%
% Private	15.6%	28.9%	38.8%	27.8%
% in North East	21.9%	17.8%	27.6%	22.5%
% in Midwest	31.3%	27.8%	23.5%	27.5%
% in South Atlantic	10.4%	18.9%	16.3%	15.1%
% South Central	22.9%	16.7%	8.2%	15.8%
% in West	13.5%	18.9%	24.5%	19.0%
% Assistant Professors	15.6%	16.1%	13.8%	15.1%
Mean Full-Time Equivalent	16.4	19.9	24.9	20.4
Mean Number of Student Activities	15.3	15.8	16.7	15.9
% with First Year Support	91.4%	97.1%	98.9%	95.8%
Mean Number of PhD Graduates	4.3	7.9	16.7	9.7

How is diversity related to the rankings of chemistry and chemical engineering programs in research universities, net of other factors associated with program rankings? Table 2 presents an Ordinary Least Squares (OLS) regression model and the beta weights predicting department rankings with diversity net of other correlates of departmental rankings. The relationship between diversity and program rankings is statistically significant at $p < .01$. With each unit increase in diversity, program rankings increase by .068 units. This relationship is net of the number of publications, citations per publication, grants, whether the institution is public or private, region, percent assistant professors, number of faculty, number of graduate student activities, percentage of first-year students receiving support, and number of PhDs produced annually.

Table 2. Regression Models Predicting Program Rankings with Diversity, Net of Other Factors

	Model I	Model II
Independent Variables	Coefficient	Beta Weights (β)
Diversity	.068[b]	.068[b]
Number of Publications	11.530[c]	.503[c]
Citations per Publication	9.761[c]	.294[c]
Grants	59.19[c]	.306[c]
Private	3.242	.050
Midwest	1.293	.020
South Atlantic	.505	.006
South Central	2.562	.032
West	2.602	.034
Assistant Professors	-11.310	-.036
FTE	.205	.075
Student Activities	.867[a]	.061[a]
First Year Support	19.86[c]	.093[c]
PhD Graduates	.040	.012
Constant	-87.54[c]	
N	278	278
R^2	.864[c]	.864[c]

[a] $p < .05$. [b] $p < .01$. [c] $p < .001$.

The beta weights—i.e., standardized coefficients—provide information about the relative strength of the relationships of the variables to program rankings. The larger the beta weight, the stronger the relationship. The standardized coefficients for this model (reported in Model II) show that the relationship between diversity and program rankings (Beta = .068) is stronger than whether the institution is public or private, percentage assistant professors, number of faculty, and number of PhDs produced annually. Therefore, not only is diversity significantly related to program rankings, but it is among its most important predictors.

Overall, the multivariate analysis strongly supports the idea that diversity is important to department rankings in chemistry and chemical engineering programs in research universities. The significant relationship between diversity and department rankings remains even after controlling for other important factors such as program faculty size, type of control, productivity indicators, and levels of student support. Combined, these factors account for more than 85% of the variance in department rankings in chemistry.

Discussion and Conclusions

This work examined the relationship between racial and gender diversity and program rankings in chemistry at research universities. The value-in-diversity perspective argues that diversity pays. It argues that diversity, relative to homogeneity, produces better outcomes for organizations. Diversity is, thus, good for educational institutions because it offers better rankings. In contrast, the diversity as process-loss perspective is skeptical of the benefits of diversity and argues that diversity is counterproductive. This view emphasizes that, in addition to dividing the nation, diversity introduces conflict and other problems that detract from an organization's efficacy. In short, this view suggests that diversity impedes group functioning and will have negative effects on performance.

Using data from the 2011 National Academy of Sciences Rankings of U.S. Research Universities, the analysis examined whether racial and gender diversity "pay" in terms of the rankings of chemistry and chemical engineering programs at research universities. The analysis centered on expectations consistent with the value-in-diversity perspective. The results were clearly counter to the expectations of skeptics who believe that diversity (and any effort to achieve it) is *harmful* to organizations. To the contrary, the results were consistent with the value-in-diversity arguments that diversity is good for chemistry and chemical engineering departments in research universities. Statistical models helped rule out alternative and potentially spurious explanations.

How is diversity related to the bottom-line performance of chemistry and chemical engineering departments in research universities? Critics assert that diversity is linked to lower quality and performance. However, the results from this analysis showed a *positive* relationship between the racial and gender diversity of chemistry and chemical engineering departments in research universities and their rankings. It is likely that diversity produces positive outcomes over homogeneity because growth and innovation depend on people from various backgrounds working together and capitalizing on their differences.

References

1. Hubbard, E. E. *Measuring Diversity Results*; Global Insights: 1997.
2. Cox, T. *Cultural diversity in organizations: Theory, research and practice*; Berrett-Koehler Store: 1994.
3. Cox, T., Jr.; Beale, R. L. *Developing Competency to Manage Diversity: Readings, Cases, and Activities*; ERIC: 1997.
4. Hubbard, E. E.; Gillium, D.; Design, K. *How to Calculate Diversity Return on Investment*; Global Insights: Petaluma, CA 1999
5. Berrey, E. C. Why diversity became orthodox in higher education, and how it changed the meaning of race on campus. *Crit. Sociol.* **2011**, *37*, 573–596.

6. Bollinger, L. Why Diversity Matters. *Education Digest: Essential Readings Condensed for Quick Review* **2007**, *73*, 26–29.
7. Gurin, P.; Nagda, B. R. A.; Lopez, G. E. The benefits of diversity in education for democratic citizenship. *J. Soc. Issues* **2004**, *60*, 17–34.
8. Hazelkorn, E. *Rankings and the Reshaping of Higher Education: the Battle for World Wide Excellence*; Palgrave MacMillan: 2011.
9. Valcik, N. A.; Scruton, K. E.; Murchison, S. B.; Benavides, T. J.; Jordan, T.; Stigdon, A. D.; Olszewski, A. M. Benchmarking Tier-One Universities: "Keeping Up with the Joneses". *New Dir. Institutional Res.* **2012**, 69–92.
10. Espeland, W.; Sauder, M. Rating the rankings. *Contexts* **2009**, *8*, 16–21.
11. Espeland, W. N.; Sauder, M. Rankings and Reactivity: How Public Measures Recreate Social Worlds. *Am. J. Sociol.* **2007**, *113*, 1–40.
12. National Academy of Sciences. Ostriker, J.; Charlotte, V.; Voytuk, J. *A Data-Based Assessment of Research-Doctorate Programs in the United States*; National Academy of Sciences: Washington, DC: 2011.
13. Baldi, S. Departmental quality ratings and visibility: The advantages of size and age. *Am. Sociologist* **1997**, *28*, 89–101.
14. Herring, C. Does diversity pay?: Race, gender, and the business case for diversity. *Am. Sociol. Rev.* **2009**, *74*, 208–224.
15. Alon, S.; Tienda, M. Diversity, opportunity, and the shifting meritocracy in higher education. *Am. Sociol. Rev.* **2007**, *72*, 487–511.
16. Iverson, S. V. Camouflaging power and privilege: A critical race analysis of university diversity policies. *Educ. Admin. Q.* **2007**, *43*, 586–611.
17. Skerry, P. Beyond Sushiology Does Diveristy Work? *Brookings Rev.* **2002**, *20*, 20.
18. Rothman, S.; Lipset, S. M.; Nevitte, N. Does enrollment diversity improve university education? *Int. J. Public Opinion Res.* **2003**, *15*, 8–26.
19. Rothman, S.; Lipset, S. M.; Nevitte, N. Racial Diversity Reconsidered. *Public Interest* **2003**, *151*, 25–38.
20. Tsui, A. S.; Egan, T. D.; O'Reilly, C. A. Being different: Relational demography and organizational attachment. *Admin. Sci. Q.* **1992**, *37*, 549–579.
21. Whitaker, W. A. *White Male Applicant: An Affirmative Action Expose*; Apropos Press: 1996.
22. Pelled, L. H.; Eisenhardt, K. M.; Xin, K. R. Exploring the black box: An analysis of work group diversity, conflict and performance. *Admin. Sci. Q.* **1999**, *44*, 1–28.
23. Pelled, L. H. Demographic diversity, conflict, and work group outcomes: An intervening process theory. *Org. Sci.* **1996**, *7*, 615–631.
24. Jehn, K. A.; Northcraft, G. B.; Neale, M. A. Why differences make a difference: A field study of diversity, conflict and performance in workgroups. *Admin. Sci. Q.* **1999**, *44*, 741–763.
25. Wood, P. Diversity in America. *Society* **2003**, *40*, 60–67.
26. Williams, K. Y.; O'Reilly, C. A. Demography and diversity in organizations: A review of 40 years of research. *Res. Org. Behav.* **1998**, *20*, 77–140.
27. Herring, C.; Henderson, L. From Affirmative Action to Diversity: Toward a Critical Diversity Perspective. *Crit. Sociol.* **2012**, *38*, 629–643.

Chapter 21

Promoting Diversity and Inclusivity at Rose-Hulman Institute of Technology

Luanne Tilstra*

Professor of Chemistry, Director of the Center for Diversity, Rose-Hulman Institute of Technology, 5500 Wabash Avenue, Terre Haute, Indiana 47803
*E-mail: luanne.tilstra@rose-hulman.edu

Rose-Hulman Institute of Technology is a primarily undergraduate school of science, engineering and mathematics. During the 2011-2012 academic year, the school defined—as one of its major goals—to "…be a diverse, globally-connected, sought-after community in which to live, learn, and work." To that end, the Center for Diversity has focused attention on growing a climate of respect for differences. Activities to promote inclusivity fall into three categories: actively maintain administrative support, entertain, and educate. Educational efforts are built on a framework defined by the Developmental Model of Intercultural Sensitivity. The success of these efforts will be monitored through responses to questions on the National Survey of Student Engagement, the Educational Benchmark Institute survey, and a modified version of the Personal and Social Responsibility Index created by the American Association of Colleges and Universities.

"The pure and simple truth is rarely pure and never simple." Oscar Wilde

The field of diversity has evolved. In the early days of diversity work, protagonists focused their energy on assuring that no individual with the desire and ability to do a particular job would be prevented from doing that job because of their gender, the color of their skin, or both. When the word 'diversity' was spoken, the words 'affirmative action' were heard (*1*). While legislation was effective at forcing diversity, folks began to recognize the need to move

© 2014 American Chemical Society

beyond 'counting heads' toward 'making heads count'. The field grew to become 'Diversity and Inclusion' as more and more people recognized the importance of welcoming the diverse perspectives and opinions that accompany colleagues from a variety of backgrounds.

Inclusion is the primary topic of this chapter. Working to increase representation of women and people of color in the sciences continues to be very important; however, it is at least as important to assure that individuals with aptitude and interest in the sciences choose to stay in the sciences. If an individual from an under-represented minority must conclude that the only way to succeed in the fields of science, technology, engineering and mathematics is to adopt behaviors and attitudes that clone those of everyone else, then the potential for true creativity is dramatically diminished; everyone loses. An inclusive environment assures that individuals with diverse perspectives have the opportunity to explore and expand their unique approaches.

The intent of this chapter is to present a framework for the development of a climate that is truly open to difference and therefore ready to accept the diverse approaches and ideas that lead to innovative and creative solutions. The chapter will describe the theory as well as the strategies adopted by Rose-Hulman Institute of Technology to put the premises of the theory into action. The ultimate goal of the ongoing work is to enhance the climate of respect for diversity.

The Setting

Rose-Hulman Institute of Technology is a primarily undergraduate institution that offers degrees in science, engineering, and mathematics. With an enrollment of 2,100 undergraduate students, the college takes great pride in its emphasis on quality education. The mission statement, "...to provide our students with the world's best undergraduate science, engineering, and mathematics education in an environment of individual attention and support," is manifested in both the attitudes of faculty and staff and in external accolades, such as a fifteen-year run with US News & World Reports 'Number 1' ranking.

Founded in 1874, the college became co-educational in 1995. Despite a delayed start allowing enrollment of women, the college has consistently matched the national average for percentage of women enrolled in all fields of science, engineering and mathematics. However, the college continues to lag behind national averages for representation of other under-represented minorities: most notably African American, Hispanic, and Native American students. Having made great strides in establishing itself as a college of choice for undergraduate engineering education, Rose-Hulman strives to be the college of choice for students from underrepresented minorities desiring an undergraduate engineering education. One approach toward accomplishing this goal is to put significant energy into the establishment of a climate of respect for diversity.

In the book 'Diversity's Promise for Higher Education', Daryl Smith (2) summarizes studies of institutions known for their ability to retain students from underrepresented minorities. Combining results from several national surveys—such as Project DEEP (Documenting Effective Educational Practice), the National Survey of Student Engagement (NSSE), and its parallel surveys for community colleges (CCSSE), law students (LSSSE) and faculty (FSSE)—Smith presents six attributes of institutions that yield success for a diverse group of students.

- The school has a mission linked to student learning and student success.
- The campus environment focuses on student learning.
- The institution favors and supports cooperative learning and collaborative activities.
- Advising and peer programs provide information about negotiating non-curricular elements.
- Faculty and student engagement in educationally purposeful activities promotes high expectations.
- The responsibility for student success is shared among all constituencies.

Because these elements closely align with the guiding principles of Rose-Hulman, the college is well-situated to incubate a novel approach toward promoting an inclusive environment. Lessons learned in this setting will provide useful starting points for a variety of educational settings.

The Role of the Center for Diversity

The Center for Diversity at Rose-Hulman Institute of Technology was established in the spring of 2011. During the 2011—2012 academic year, a Diversity Steering Committee comprised of eight faculty and staff met nine times in order to define a strategic plan. The purpose of the plan was to prioritize the tasks and define a vision for the future of diversity at Rose-Hulman. Following is the vision statement adopted by the Diversity Steering Committee and approved by the Cabinet.

Rose-Hulman Institute of Technology is committed to building an inclusive community in which the multiplicity of values, beliefs, intellectual viewpoints, and cultural perspectives enrich learning and inform scholarship.

To fulfill this vision, three goals were proposed.

I. Make diversity an integral component of campus life.
II. Foster respectful relationships among people of different cultures, lifestyles, and perspectives.
III. Expand awareness among local and global constituencies of Rose-Hulman's activities to enhance diversity and promote inclusivity.

Specific activities and tasks are part of each goal; tasks or activities that do not fit within these goals are filed as items to be considered for the next strategic plan. The tasks under the first goal focus on establishing and maintaining administrative support. Enlisting and maintaining administrative support is identified as being of primary importance and represents the first of three avenues in the action plan.

Activities under the second and third goals include hosting cultural festivals and other largely visible events; these activities (which usually involve food) are well-attended. Visible educational and celebratory events represent the second of three avenues in the action plan. While enjoyable and popular, the trifecta of 'Fun, Food, and Festivals' does little to engender self-reflection or attitude-changing experiences. For this reason, accomplishment of the second and third goals must also include some sort of focused training.

Guest speakers, workshops and other deliberate events are the third avenue of the action plan. However, rather than creating a series of workshops with random, disconnected topics, the decision has been made to create a series of workshops that have a defined goal. The nature of that goal is defined by a theory developed by those who study interactions between cultures that exist in different countries. The theory posits that an ability to appreciate different cultures requires being able to communicate competently with those cultures, and that an ability to communicate competently with different cultures requires having intercultural sensitivity. Consequently, it is the goal of diversity training at Rose-Hulman Institute of Technology to increase the intercultural sensitivity of the campus by building intercultural competencies in the stages defined by the Developmental Model of Intercultural Sensitivity. It is believed that increased intercultural sensitivity will yield a more inclusive campus climate.

The Developmental Model of Intercultural Sensitivity

The Developmental Model of Intercultural Sensitivity was first presented by Milton J. Bennett (*3*). A core premise of this theory is that an ability to have productive interaction with individuals from cultures that are significantly different than one's own is acquired gradually and continuously upon exposure to difference. Although the work was initially presented to individuals working in international settings, it requires no great stretch to see its value for helping individuals trying to understand different cultures within our one, culturally diverse nation.

The Developmental Model of Intercultural Sensitivity (DMIS) presents six levels of awareness. The model proposes that as individuals are exposed to cultural differences, their response to the differences will go through stages. The model also proposes that individuals who are aware of these stages are more likely to move through the stages productively.

The first three stages of the theory are ethnocentric; that is, individuals in the first three stages are inclined to view cultures through the perspective of their own experience. The stages include denial, defense, and minimization. Individuals in the first stage, ***denial of difference***, are unable to construe differences between cultures. This stage is benign on the surface, but includes a tendency to dehumanize outsiders. Folks in denial of difference may attribute inferior intellect to individuals who present behaviors that deviate from what is expected. Because of this, folks in denial are not troubled by the practice of exploitation of cultures that differ from their own.

Individuals in the second stage, ***defense against difference***, recognize the existence of cultural difference and consider difference in a negative light. Folks in this stage see one culture as being superior to all others; they tend to see neutral or positive statements about any other culture as an attack on the culture they prefer. Interestingly, individuals in denial may select their own culture as the superior culture or they may select some other culture as being superior while maligning their native culture. A common feature of folks in the defense of difference stage is dualistic thinking. Folks in this stage feel totally justified denying equal opportunity; e.g., '____ are all naturally passive, hence they are all poor leaders.'

Individuals in the third stage, ***minimization of difference***, recognize the existence of cultural differences while holding that all human beings are essentially the same. While folks at this stage are consistently and insistently nice, they may not notice inherent differences that rise out of different genetics and different cultures. They tend to believe that—deep down—everyone is the same. They accept outward manifestations of difference but deny the existence of genuine 'other'. Consequently, folks who minimize cultural differences are not aware of the possibility of viewing the world through another's eyes.

The next three levels of the theory are ethno-relative; that is, individuals are able to recognize the value of switching their perspective from one culture to another, depending on the current set of experiences. The stages include acceptance, adaptation and integration. Folks in the fourth stage, ***acceptance of difference***, accept cultural differences as viable alternative solutions to the organization of human existence; whether beliefs, behaviors and values are 'good' or 'bad' depends on the cultural context. The fifth stage, ***adaptation to difference***, includes an effective use of empathy and the ability to shift one's frame of reference in order to understand and be understood across cultural boundaries. Individuals in the sixth stage, ***integration of difference***, have internalized multicultural frames of reference to the extent that they seamlessly move between ways of being. They are often unaware of their own transitions.

DMIS Informs Curriculum Development

Clarifying the aspects of these six stages introduces an explanation for why two individuals—when presented with the same experience—may present totally different responses. Anyone who has ever taught General Chemistry recognizes the value of presenting concepts in a particular order. Trying to teach the concept of dynamic equilibrium to someone who has not mastered an understanding of the difference between an atom and a molecule leads to confusion on the part of the student and frustration on the part of the teacher. However, worse than the lack of learning is the very real possibility that the student will become convinced that chemistry is illogical and cannot be learned. In the same way, a workshop participant who is in the second stage of intercultural sensitivity (defense) is unlikely to learn from a video that emphasizes the ways a dominant culture has hurt non-dominant individuals; rather the individual is very likely to become convinced of their own 'rightness'. It is very possible that—like the frustrated chemistry student—the workshop participant will become more actively engaged in 'proving that the dominant culture is the best culture. The individual may well become more resistant to moving beyond the attitudes of the second stage (defense).

As is often the case, theory can be used to propose a method to overcome the challenges of a task. For this reason, the educational experiences designed to enhance inclusivity at Rose-Hulman will be based on the hypothesis that workshops designed to build competency in the order presented by the theory of DMIS will have a positive impact on the climate of respect for diversity.

Content and Format of the Workshops

During the summer of 2013, twenty-four faculty and staff were invited to participate in three days of intensive training the purpose of which was to expose attendees to the six levels of DMIS in a way that would help them experience personal growth through a series of activities. At the conclusion of the three days, participants helped to define the curricula for similar workshops to be offered to the rest of the faculty and staff. First offered during the 2013—2014 academic year, sessions have been facilitated by individuals selected from the team of twenty-four.

It is the premise of this approach that when respectful behavior is modelled by the faculty and staff, students will respond positively to similar instruction. Therefore, a subset of this team will create sessions specifically designed for students to be presented beginning in the fall of 2014.

Attempts to promote diversity benefit from strategies that avoid pitting group against group and are welcoming to everyone. For this reason, early workshops are designed to be inclusive and non-threatening. The intent of the first workshop is to arouse curiosity while establishing a vision of constructive relationships. As shown in Table I, activities in the first workshop involve guiding participants through exercises that help them increase awareness of their own culture as well as an awareness of differences in communication styles. Using the DMIS theory as a framework, workshops are designed to help folks grow into the next level. Participants are required to complete the workshops in order so that they may be challenged to grow at a level that matches their current stage of intercultural sensitivity.

Response to Workshops

At this point in time, the first workshop has been offered five times and the second workshop has been offered twice; enrollment is voluntary. Rose-Hulman has approximately 170 faculty and 330 staff. Seventy-five individuals have participated in the first workshop and—so far—twenty of the seventy-five have participated in the second workshop. In both workshops, one-third of the participants have been faculty. The sessions have been well-received. Participants enjoyed the opportunity to interact with individuals from other departments and made comments about recognizing the different communication styles among colleagues.

Table I. A summary of important considerations when designing workshops for Intercultural Communication Competence

	Activities must ...	Learning Outcomes
Denial of difference		
	Be user friendly Be non-threatening Be inclusive, non-blaming Around curiosity	Definition of culture Increased cultural self-awareness Understand different verbal communication styles
Defense of difference		
	Avoid cultural contrasts Promote cooperative activities Address feelings & emotions Develop cultural self-awareness	Identification of own strengths, needs, values Recognition of other strengths, needs, values Skills to handle differences
Minimization of difference		
	Avoid pushing Provide complex frameworks for comparison Structure opportunities for difference-seeking	Recognize that beliefs, attitudes and values exist in a cultural context Dimensions of Diversity Recognize differences in non-verbal communication patterns
Acceptance of difference		
	Prepare for frame of reference shifting Promote examination of more profound contrasts Motivate complete cultural observations	Awareness of one's own frame of reference Curiosity about what happens 'inside' other cultures Recognize privilege within one's own culture
Adaptation to difference		
	Address deeper anxiety issues Facilitate opportunities to practice behaviors Facilitate multicultural group discussions	Ability to adapt communication patterns Interaction management skills Problem-solving skills
Integration of difference		
	Provide options for individuals to be a resource Develop a peer group	Ability to create new categories Flexibility, empathy

Plans for Assessment of Campus Climate

Rose-Hulman has historically participated in several surveys to monitor aspects of campus climate. Respect for diversity is one of the aspects for two of the surveys administered to students; specifically the National Survey of Student Engagement (NSSE) and the Educational Benchmark Institute (EBI) survey. Monitoring responses to questions in these surveys that are relevant to campus climate will continue. In addition, a modified version of the Personal and Social Responsibility Index (PSRI) created by the American Association of Colleges and Universities was administered to all faculty, staff, and students during the first two weeks of December, 2013. The PSRI was modified to add questions that specifically probe the climate of respect for diversity; results will be available in January of 2014. The survey will be repeated in three years in order to determine the impact of this work on the campus climate.

Rose-Hulman Institute of Technology is in the early stages of developing a program to enhance diversity and promote inclusivity on campus. The Center for Diversity has been spear-heading these efforts by first completing the study necessary to design a strategic plan, then using that Strategic Plan for Diversity to create an action plan. Activities to promote inclusivity fall into three categories: actively maintain administrative support, entertain, and educate. Educational efforts are built on a framework defined by the Developmental Model of Intercultural Sensitivity. Three workshops have been designed based on the tenets of DMIS; two have been offered and were well-received by administrators, faculty, and staff. Three more workshops that continue the curriculum will be designed over the next year. Individuals interested in more detailed information about specific workshops are invited to contact the author.

References

1. Gardenswartz, L.; Rowe, A. *Managing Diversity: A Complete Desk Reference & Planning Guide*; The Society for Human Resource Management: Alexandria, VA, 2010.
2. Smith, D. G. *Diversity's Promise for Higher Education*; The Johns Hopkins University Press: Baltimore, MD, 2009.
3. Bennett, M. J. In *Education for the Intercultural Experience*, 2nd ed.; Intercultural Press: Yarmouth, ME, 1993.

Chapter 22

Improving Transparency and Equity in Scholarly Recognition by Scientific Societies

Erin L. Cadwalader[1] and Amanda C. Bryant-Friedrich*,[2]

[1]Association for Women in Science, 1321 Duke St., Suite 210, Alexandria, Virginia 22314
[2]Department of Medicinal and Biological Chemistry, The University of Toledo, 2801 W. Bancroft St., MS#606, Toledo, Ohio 43606-3390
*E-mail: Amanda.Bryant-Friedrich@utoledo.edu

Anecdotally, it had been noted that women were underrepresented among scholarly award recipients from technical disciplinary societies relative to the available pool, but overly recognized relative to that same pool for service and mentoring awards. Evaluating the award recipients for several dozen societies in a wide range of disciplines revealed this is indeed the case, regardless of the field of study. In an effort to bring greater transparency to the awards process, the Association for Women in Science partnered with 18 scientific technical disciplinary societies, including the American Chemical Society to increase the transparency and equity of the awards process. The following chapter discusses the best practices and outcomes resulting from this collaborative partnership.

Introduction

The Association for Women in Science (AWIS) is the largest multi-disciplinary organization for women in science, technology, engineering and mathematics (STEM) in the United States. This organization, founded in 1971, is dedicated to achieving equity and full participation of women across all employment sectors within the STEM disciplines. In its efforts to achieve this goal, AWIS establishes collaborations with various players in the STEM community to gain an understanding of discipline-specific challenges and to

© 2014 American Chemical Society

provide strategies to eradicate many of the significant issues which still hamper the progression of women in the sciences, mathematics, engineering and technology. As a part of these efforts, AWIS has developed a relationship with the American Chemical Society (ACS) to explore the nature of the scholarly awards process in the chemical sciences and to develop strategies to increase the numbers of citations awarded to women in this discipline.

Women have dramatically increased as a percentage of the chemistry community in the last several decades. In 2010 alone, according to data collected by the National Science Foundation (NSF), women earned 50% of the bachelor's degrees and 39% of the PhDs in chemistry, and comprised 36% of employed chemists across all degree levels. In 1982, women earned 16% of the PhDs granted in chemistry and 30 years later, 14% of the tenured faculty positions in chemistry were earned by women. However, in the decade leading up to 2010, women on average won only 7% of the nearly 50 scholarly awards which are granted by the ACS. In contrast, the membership of the ACS was comprised of 21% women during the same time period.

Why do awards matter to individuals? Scholarly awards are given to recognize distinguished accomplishments in one's field and career. They are important at various stages of one's professional trajectory for promotions, for positive tenure decisions as well as for obtaining leadership positions in one's institution or academic community. Not insignificantly, they provide a sense of acknowledgement and appreciation by one's professional community. Management studies have demonstrated that most people would rather be publicly recognized for their achievements and hard work than receive a raise (*1*). With some awards, you receive both!

Technical disciplinary societies (hereafter referred to as Societies) give awards for a variety of reasons. They recognize pioneers and leaders in the discipline, who serve as role models for aspiring scientists. They may also highlight cutting edge research and emerging fields. Both of these reasons motivate and inspire scientists, especially those at the early career stage. Societies typically give several different types of awards, including scholarly awards to recognize senior researchers and lifetime achievements. Junior investigator awards are given to rising talent in the field. Awards for service, mentoring, and teaching recognize participation in the community and molding of the next generation.

Additionally, the Society may have been given money to endow an award. More than half of the scholarly awards the American Chemical Society confers are named after individuals. The Alfred Bader Award in Bioinorganic or Bioorganic Chemistry, the Alfred Burger Award in Medicinal Chemistry, the Arthur C. Cope Award, and the Claude S. Hudson Award in Carbohydrate Chemistry all share several things in common. The names all represent pioneers in their respective fields in chemistry. These individuals represent important accomplishments which form the foundation for the science. They are also all named after white men. While named awards are a great way to honor pioneers, they may also send the message that the upper echelons of the society remain dominated by white males. This has a subtle but important effect on those being evaluated to win the award, those thinking of whom to nominate for it, and those who may look at the awards and realize they don't see anyone who looks like them reflected in the composition of elite players in the discipline or organization. This is one example of several considerations Societies and all other award giving groups need to consider when trying to improve the equity and transparency of their awards processes.

Many professional societies also offer women-only or minority-related awards. Research indicates that when societies give women-only awards, women are even less likely to win scholarly awards. This serves to set women's research apart in a way that is antithetical to the goal of not viewing research done by women as a thing apart or separate and substandard to that of their male contempories. It is for this reason that AWIS chose to exclude this type of award in this study.

Methods

Former AWIS president, Dr. Phoebe S. Leboy, observed that women seemed to win fewer awards for their scholarly achievements from technical disciplinary societies than would be expected based on the available pool, while they seemed to be overly recognized for mentoring, teaching, and service. Dr. Leboy commenced a study to examine whether this was merely a perception or a statistical reality.

She found that in the decade leading up to 2010, women on average won only 7% of the nearly 50 scholarly awards bestowed by the ACS (Table 1), despite making up 14% of tenured faculty at the top 50 research institutions. In addition, ACS has enjoyed 15% female leadership in the office of president since 1990. In contrast, 38% of the education and teaching and 18% of the service awards for the same time frame went to women (Table 2). This disparity is, however, not restricted to chemists or to the ACS. Regrettably, this phenomenon was exhibited across all disciplines of STEM (*2*).

Table 1. Percentage of scholarly awards presented to women

	Total Scholarly Awards	*Scholarly Awards Won by Women*	
Year(s)	N	N	%
1991-2000	391	10	**2.6%**
2001-2010	440	32	**7%**
2011	48	7	**15%**
2012	47	3	**6.4%**
2013	50	6	**12.0%**

Table 2. Percentage of teaching and service awards presented to women

	Total Ed. & Teaching Awards	Education & Teaching Awards Won by Women		Total Service Awards	Service Awards Won by Women	
Year(s)	N	N	%	N	N	%
1991-2000	56	9	16.1%	32	4	12.5%
2001-2010	64	24	37.5%	44	8	18.2%
2011	12	4	33.3%	4	1	25.0%
2012	7	4	57.1%	4	1	25.0%
2013	7	2	28.6%	5	3	60.0%

To change this troubling trend, AWIS initiated several collaborations with Societies who suffered from the same outcomes regarding women and scholarly awards. The ACS was one of the earliest society partners. A grant from the National Science Foundation, **Advancing Ways of Awarding Recognition in Disciplinary Societies** (AWARDS, ADVANCE Grant #0930073) helped facilitate the process.

In 2010, a workshop brought these society partners together to discuss best practices for the awards process. They were introduced to concepts such as implicit bias, gender neutral language, and establishing evaluation criteria. Each society worked with AWIS to develop an action plan, which is discussed in the Results section. The representatives then took these plans back to their Societies for discussion and implementation. In 2012, a follow up workshop was hosted to bring in the original eight Societies to discuss the advances they had made as well as the challenges they had overcome, and to share their experiences with eleven new Societies who also sought to improve the transparency and equity of their own awards processes. AWIS has continued to track the outcomes for ACS, as well as the other Societies, as they move forward with their implementation plans.

Results

Following the 2010 workshop, the leadership representing ACS returned to their society governance with an action plan which was presented to the ACS Board Committee on Grants & Awards and the Joint Subcommittee on Diversity during their meetings that summer. The specific points of action are discussed below, with explanations about the relevance of each topic.

Inoculation against Implicit Bias

Scientists pride themselves on objective evaluation of outcomes, and indeed it is the bedrock of the scientific method. We like to believe that we evaluate everything based on the best available data, applying reason free from passion. Science, however, demonstrates that we are human at the core and thus experiences and the cultural influences in which we were raised impact our perceptions. A recent study examined how scientists in multiple disciplines, including chemistry, evaluated two candidates for a lab manager position intended for someone who recently finished their bachelor's degree (3). The only thing that varied between the two candidates among otherwise identical application packages was their name; one was indentified as a woman and the other as a man. Both men and women evaluating the candidates gave higher scores to the man in terms of competence, willingness to hire and mentor the individual, and the salary they would offer.

As there was nothing different between the candidates besides gender, it is clear that both women and men have biases which make them more likely to associate men with being better at science. This response, an unconscious judgment based on the culture in which one is raised, is called unconscious or implicit bias (4, 5). It is seen in many different variations depending on the culture in which one is raised and in other professions as well. For example, women have historically been perceived as sub-par musicians compared to men, but when blind auditions were conducted with a screen separating the musician from the reviewers, the number of women in orchestras increased (6). In a Harvard Business School Class, students were given one of two identical case studies documenting the stellar career of one individual, where again only the name was changed, and then students were asked to evaluate the candidate for various qualities (7). They found the male and female candidate to be equally competent, but were much more severe in judging the woman candidate's personality. They found "her" aggressive nature off-putting and felt "she" was selfish, demonstrating that success negatively correlates with likeability for women.

Unfortunately, the influence of these subconscious biases have a profound impact on outcomes in the scientific community as well. A study of grants given by the Swedish Medical Research Council found that women had to perform at a substantially higher level than men to receive equal levels of funding (8). Another study found that double blind reviews of submissions to an ecology journal, where neither the names of the reviewers nor the names of the authors are known to each other, increased the number of female authors (9). Male and female psychology professors evaluating "applicants" for an academic psychology position, using CVs that varied only by the gender of the applicant, found the evaluators more positively valued the male applicant's research, teaching, and service over the female's (10). Furthermore, they felt the male applicant was better suited for the associate professor level while the female should be an assistant professor. Children asked to draw a scientist routinely produce images of a white male with wild hair, indicating that these biases are deeply ingrained, which makes them difficult to combat (11). However, the effects of biases can be partially mitigated by making people aware of them (12). The importance of educating people about their biases applies not only to evaluations of women but also to discussion about implicit biases related to racial and ethnic diversity.

To that end, the ACS Action Plan involved two points committed to reducing the influence of implicit bias on the awards process. ACS planned to develop a webinar on implicit bias to be viewed by all members of canvassing and selection committees. Viewing these webinars would be an important part of committee training prior to committee deliberations. Furthermore, ACS planned to make attendance at the webinar a part of the dedicated responsibilities of the chair of each selection committee. As both men and women hold these biases, it is important that all participants in the process are educated about the influences they impart.

Review and Update Awards Portfolio

The names of awards can have a broader impact than intended on the pool of nominees and subsequent winners of society prizes. One thing Societies should consider when reviewing the portfolio of awards is whether they are all named after pioneers in the field. Due to the history of inequality in the scientific disciplines this group is strongly lacking in diversity. Do these awards continue to reflect the diversity of the discipline? Do the images of these individuals have an impact upon public perception of who the award should be given to because they assume subconsciously that the awardee should bear some resemblance to those for whom the award is named? This should not be a consideration when evaluating who will win the award, nor who should be nominated or considered for nomination, but we know subconscious association has a strong impact and this is a consideration that should not be overlooked as it may discourage the nomination of perfectly good candidates.

Another consideration when evaluating the portfolio of awards is whether each award reflects a sub discipline that is still relevant. Have new fields emerged which aren't being recognized? What about interdisciplinary specialties, a space which tends to attract more women than men? The number of nominees for an award should be an indication of the strength of some sub disciplines relative to others. Giving consideration to the number of nominees may also impact the timing of awards and the frequency with which they are given. If a field is shrinking, a suggestion is to consider granting the award every two or three years instead of annually. Also consider what the diversity of nominees looks like for these awards from year to year. Are the nominees all men? Consider what this may say about the award and the sub discipline.

To that end, there are multiple factors which can be improved when increasing the transparency and equity of the awards process. Evaluating the language used to describe the qualifications for each award is important. Clearly delineated criteria which must be met in order to be eligible for the award and on which grounds the winner will be decided, should be free of gender bias. Also worth considering are age limits placed on eligibility as women sometimes take time off and then return to the work place, disqualifying them based on age for awards for which they should be eligible based on training.

The ACS Action Plan involved several objectives related to the considerations listed above. The committee planned to discuss areas for new awards and the appropriate considerations which should be taken in the process of creating them. ACS planned to examine their current national awards and look for bias in the purpose and eligibility statements, as well as to review the age limits for awards and see if they were discriminating against potential candidates that took a less than traditional route or took time off at any stage in their training. As a way of evaluating the outcomes of these efforts, they developed a strategy to target awards which experienced the lowest numbers of female nominees and recipients and focus specific efforts around these initiatives.

Develop a Diverse Pool of Candidates

One of the biggest challenges to increasing the equity of the awards process is simply increasing the number of nominees. People assume that there are lots of nominees for each award, but often that is not the case. There are a variety of ways to increase the nominee pool, however. One of the most important ways is to simply make sure the solicitations are broadly distributed. This means using technology to disseminate them, such as social media and email, but also using print versions in magazines and newsletter and encouraging members to post them in common spaces such as a bulletin board dedicated to such announcements.

Encouraging individuals to submit nominations is another approach to diversifying the candidate pool. Department chairs, scientists within the National Academy of Sciences, deans of science colleges, leaders of technical divisions in industry, and members in society leadership positions are all individuals likely to know qualified candidates but may not think to make a nomination without some prompting. Creating a society-wide nomination or canvassing committee to specifically look for those who have displayed excellence in their achievements but may not have strong advocates in their immediate community is another approach. This committee should collaborate with other entities within the organization which have a history of supporting the nomination of women for awards such as a women's committee and/or a diversity committee.

Clearly defining the award criteria as well as the requirements for a complete nomination package, and making sure this information is easily available is another way to lead to an increased number of nominees. Posting criteria and requirements in a prominent part of the society webpage is important so that one doesn't have to hunt to find the information. Part of that also means using gender neutral language in the solicitations. Certain words are subconsciously associated with masculine traits, such as *exceptional, outstanding, brilliant*, and *analytical* while words with feminine associations include *thorough, conscientious*, and *methodical* (*13*). It is important to take steps to make sure these subconscious cues don't negatively influence potential nominators as well as potential candidates. Using gender neutral language is important not only for solicitations, but also in letters of nomination, as this has been shown to be a serious problem when considering women for other leadership positions (*14*).

To address the lack of diversity in award nominations, the ACS planned to review their dissemination strategy and try to adopt one with broader distribution. They also planned to develop a list of tips on creating nomination packages, which include information on implicit bias and gender neutral language considerations. A serious consideration of the society centered around the inclusion of questions regarding how the nominator knows the nominee, why this individual is being nominated for this specific award, and how this nominee compares to others in the field. Lastly, ACS intended to make sure information regarding awards was readily available on the website.

Evaluate Nominees Objectively

While getting a diverse pool of nominees is part of the challenge of creating greater equity in the awards process, creating greater transparency is equally as important to decreasing the influence of bias and evaluating nominees objectively. Creating diverse nomination and selection committees is crucial. Studies have shown that one underrepresented member on a committee is not sufficient because they are less likely to speak up. Aiming for a better balance of genders and ethnicities on a panel is important as greater variety of perspectives leads to broader considerations and better decision making. Part of this equation is having good committee dynamics. Making sure there is sufficient time to evaluate each candidate so that each individual can share their opinion is very important. Conducting these meetings in person is best, but if that is not possible, using a conferencing option such a Skype® where all individuals can see each other is better than merely conducting the discussions over the phone.

In the previous sections, the importance of clearly defined evaluation criteria was discussed. Prior to discussing the candidates, the group should discuss the criteria being evaluated, and establish which qualities are viewed as most important for selecting the award recipient. This helps make sure the nominees have the correct qualifications and reduces cronyism. Disclosing potential conflicts of interest is also important for objective selection of award recipients. Discussing what appropriate levels of distance from the nominees are is important. Things to consider include mentoring connections, training at the same institution, being employed at the same institution, or past collaborations conducted with nominees.

The actions for the society included much consideration for past committee compositions and outcomes. For example, of the 10 awards which never had any female nominees, an examination of the composition of these committees was planned. Committing to ensuring the committees are diverse is an important step. Making sure the criteria for each award is clearly delineated, and that discussion of these metrics and agreement on priorities is part of the selection process is another crucial practice to implement. ACS specifically planned to address this with the 2010 selection committees and request that they document selection criteria that summer during meetings of the selection committees. Additionally, they planned to implement the same type of best practices for the committees that decide the newly announced (2010) ACS fellows.

Results

Increasing Awareness of Issues within the Community

ACS used the 2011 annual meeting as a platform to present information on the underrepresentation of women among nominees and awardees. Implicit bias was discussed and examples of letters with gender-neutral language were shared *via* a formalized and web-available presentation. Information regarding selection and canvassing committees, as well as selection committee best practices, is currently available on the ACS website.

Increasing the Diversity of the Nomination Pool

The plan to increase the number of nominations included increasing the exposure of solicitations to the target community by rethinking print and media dissemination strategies. An article was published in *Chemical and Engineering News* as a call to action and the website was updated to include revised and updated information. A variety of individuals in ACS leadership positions, including technical division leaders, national committee chairs, and local section officers in ACS were informed of the new resources available for award selection committees. Canvassing committees were created to help solicit nominations from others in the technical community as well. A Women Chemists Committee Task Force was also reconstituted to help with these sustained efforts.

Addressing Committee Diversity and Other Best Practices

Measures to increase the transparency of the awards process included a variety of activities. Canvassing Committees helped find nominees for the awards committees that historically received low numbers of or no female nominees. The ACS also increased the proportion of women appointed to selection committees, partly by appealing to the membership for nominations of women. More women were appointed as members and chairs of selection committees. Selection committees were encouraged to review materials about implicit bias and best practices for selection of diverse pools of candidates for awards and committees. They were also directed to emphasize the significance and impact of the research contributions of each nominee during their discussions of candidates. This drew the focus away from other factors such as scientific pedigree which often has a significant impact on selection.

After evaluating the awards from 2007-2012, they found that for all awards, not just scholarly, women were 13% of the nominees, and 18% of the awardees. For the scholarly awards, however, women were 7.7% of the nominees and 17 awards had less than 3% female nominees. In 2011, there was an initial increase in the number of female scholarly award winners, with women winning 15% of the awards. This number however decreased to 6.4% in 2012, leading the society to reinvigorate efforts, resulting in an increase to 12% in 2013 (Table 1).

Evaluation of National Academy of Sciences Influence

Committee members were surveyed after their work to identify what selection criteria were used and how the discussions proceeded; they were specifically asked about the impact of candidate membership in the National Academy of Sciences (a highly-coveted honor in the United States) on their decisions. Results of the survey indicate that members of the National Academy had an edge in being selected for awards, but nominees with letters from National Academy members were not preferentially selected for awards.

Conclusions

Women in the chemical sciences in the United States as well as around the world have experienced many challenges as they have sought equity and the opportunity to fully participate in the activities which make up the scientific enterprise. The receipt of awards from scientific and professional societies around the world is a pleasure which is far too often reserved for men. The best practices outlined in this chapter have been deeply explored by the ACS as well as other professional societies, and provide a foundation on which to build fair and transparent awards programs both nationally and internationally. In addition to the initiatives outlined here, other societies have chosen to identify networks of influence and connectivity within their organizations to determine how these groups influence the awards process. Others have chosen to adapt a double-blind review process for its journals, as publication is a major factor in the selection of awards recipients within their discipline.

AWIS will continue to monitor the progress of the ACS as well as other professional societies and to help these organizations modify and update their practices to achieve an equitable distribution of awards to all genders, races and ethnicities represented within their respective memberships. AWIS sees this as an important part of ensuring the maintenance of a globally productive and competitive scientific enterprise.

References

1. Frey, B. S. *EMR* **2007**, *4*, 6–14.
2. Lincoln, A. E.; Pincus, S. H.; Bandows Koster, J.; Leboy, P. S. *Soc. Stud. Sci.* **2012**, *42*, 307–320.
3. Moss-Racusin, C. A.; Dovidio, J. F.; Brescoll, V. L.; Graham, M. J.; Handelsman, J. *Proc. Natl. Acad. Sci. U. S. A.* **2012**, *109*, 16474–16479.
4. Greenwald, A. G.; Banaji, M. R. *Psychol. Rev.* **1995**, *102*, 4–27.
5. Greenwald, A. G.; Nosek, B. A.; Banaji, M. R. *J. Pers. Soc. Psychol.* **2003**, *85*, 197–216.
6. Goldin, C.; Rouse, C. *Am. Econ. Rev.* **2000**, *90*, 715–741.
7. Ely, R. J.; Ibarra, H.; Kolb, D. M. *Acad. Manage. Learn. Educ.* **2011**, *10*, 474–493.
8. Wenneras, C; Wold, A. *Nature* **1997**, *387*, 341–343.
9. Budden, A. E.; Tregenza, T; Aarssen, L. W.; Koricheva, J; Leimu, R; Lortie, C. J. *Trends Ecol. Evol.* **2008**, *23*, 4–6.
10. Steinpreis, R. E.; Anders, K. A.; Ritzke, D. *Sex Roles* **1999**, *41*, 509–528.
11. Chambers, D. W. *Sci. Educ.* **1983**, *67*, 255–265.
12. Abrams, D.; Crisp, R. J.; Marques, S.; Fagg, E.; Bedford, L.; Provias, D. *Psychol. Aging* **2008**, *23*, 934–939.
13. Schmader, T.; Whitehead, J.; Wysocki, V. H. *Sex Roles* **2007**, *57*, 509–514.
14. Madera, J. M.; Hebl, M. R.; Martin, R. C. *J. Appl. Psychol.* **2009**, *94*, 1591–1599.

not have a STEM degree or who are not Native may join AISES as Associate Members. Membership in AISES is open to everyone who supports the AISES mission.

AISES also has student members at all levels. AISES currently has about 3,000 members, about half of whom are professionals and half are students.

AISES is addressing the gross underrepresentation of Native People in STEM at every level, including educational access, degree attainment, and the workforce.

Why the AISES Mission?

Naturally one reason for the AISES mission is to enable Native People to benefit from the rewards, fulfillment, and thrill of a career in the sciences. Science is fun and everyone should have the opportunity to join in the excitement of discovery.

Another important purpose is the development of technical expertise and enterprise for the benefit of Native America. The ongoing trend is that some of our former AISES college students return to Native communities as technical innovators, teachers, medical professionals, environmental professionals, and technological business developers; some have taken roles in tribal governance.

Perhaps the most important rationale of all is that Native People have so much to contribute to science. Among Indigenous People there is a long tradition of observation of nature and of using knowledge acquired from careful observation of the natural world to enhance quality of life and to survive and thrive under the local conditions and climate. The intelligence, resourcefulness, talent, and creativity of Native People, and also Native values, should not be missing from our profession.

The AISES Membership

The AISES Membership, known as "The AISES Family," consists of:

- K-12 students in schools of all types – Bureau of Indian Affairs (BIA), public, private, reservation, rural, and urban
- Secondary and post-secondary educators
- Undergraduate and graduate college students
- Postdoctoral researchers
- Professionals in academia, government, industry, and the military
- Retirees
- Council of Elders

These members represent over 200 tribal nations and span all STEM fields.

AISES Programs

Pre-College Programs: Awareness and Retention

We at AISES believe that in order to develop and expand the future cadre of Native STEM professionals, it is essential to engage our youngest community members. AISES programs for pre-college students are designed to raise awareness of the STEM fields and STEM careers, to promote retention and persistence in school, and to increase science and mathematics interest and skill set (Figure 1 and http://www.aises.org/programs/pre-college). Some of this work is accomplished through the AISES Pre-College Affiliated Schools Network, to which over 150 schools belong.

Figure 1. Images from AISES pre-college programs. (Courtesy of AISES.)

Power Up workshops, presented by AISES at schools across the country, are designed to promote awareness of, and participation in, science fairs. K-12 students learn about the scientific method and how to conduct scientific research. Students are encouraged to initiate projects and to participate in local, regional, and national science fairs.

K-12 students have the opportunity to showcase their work on science projects each year at the National American Indian Science and Engineering Fair (NAISEF) (Figure 2). NAISEF allows students to show off their work to a national audience, practice oral presentation skills, and connect with other Native students who are science buffs. NAISEF winners present their projects at the International Science and Engineering Fair (ISEF); AISES covers the travel expenses to ISEF for NAISEF winners and their chaperones.

College Programs: Access and Success

The AISES college programs provide support to undergraduate and graduate students in STEM fields through college chapters, regional and national conferences, leadership development, mentorship, scholarships, internships, and career resources (http://www.aises.org/programs/college).

Many college student members of AISES are active in an AISES college chapter across the country. Currently there are 85 active AISES college chapters in good standing; a total of 184 college chapters have participated in AISES in recent years. These chapters serve as a cultural home for Native students, providing social and academic support and disseminating information about opportunities. Each chapter is led by student officers and is guided by one or more faculty advisors. Figure 3 shows the members of the award-winning AISES chapter from the University of Minnesota - Morris.

Figure 2. Projects presented by K-12 students at the National American Indian Science and Engineering Fair (NAISEF). (Courtesy of AISES.)

The AISES college chapters are divided into seven geographic regions and every year in the spring one college chapter in each of the seven regions organizes and hosts a regional conference. At the regional conferences, high school and college students, AISES professional members, and AISES partners are brought together to provide networking opportunities, informative sessions, and presentation opportunities.

Figure 3. The award-winning AISES college chapter from the University of Minnesota-Morris, at the AISES National Conference in Anchorage, Alaska, November 2012. (Courtesy of AISES.)

Natives with similar professional interests and also have the opportunity to serve as mentors to Native students.

The AISES National Conference

Every year in the fall, AISES members come together at the National Conference. The conference sessions are organized into tracks for high school students, college students, educators, and professionals and are focused on educational, career development, and cultural themes. The Career Fair at the AISES National Conference is the largest STEM career event in Indian Country and features jobs and a host of educational and internship opportunities. The Career Fair includes industry employers seeking talent as well as colleges and universities, non-profit organizations, and tribal agencies (Figure 5).

Figure 5. The Career Fair at the AISES National Conference. (Courtesy of AISES.)

Chemistry and the American Chemical Society certainly have a strong and visible presence at AISES events. The ACS Scholars Program is promoted at the Career Fair and ACS members are regular presenters at AISES National. In 2009, then ACS President Dr. Thomas H. Lane gave a plenary address at the AISES National Conference in Portland, OR and also at other AISES events. Former ACS President Dr. Nancy B. Jackson (Seneca) has given multiple plenary speeches at AISES National and at the Leadership Summit. The International Year of Chemistry was featured in a plenary session at the 2011 AISES National Conference in Minneapolis, MN. A picture of some of the ACS members at the 2012 AISES national conference is shown in Figure 6.

AISES Elements of a Successful Model for Diversity and Inclusion

Engagement of the Youngest Members of Our Community

From the time of our Founders, AISES recognized that, in order to increase participation of Indigenous People in STEM, we need to cultivate interest within our young people and promote pursuit of college degrees and advanced degrees. K-12 programs, designed to introduce children to the excitement and promise of the world of science, are a critical piece of any successful STEM diversity model.

Figure 6. Some of the ACS members in attendance at the 2012 AISES National Conference in Anchorage, AK. (Courtesy of AISES.)

Through our college chapters located all across the country, AISES has a unique pathway to bring science to K-12 Native students. Outreach is one component of our college chapters' ongoing activities and our chapters have been very successful in engaging younger students in science experiences and in giving K 12 students a glimpse of college life. For example, through a grant from the National Science Foundation Directorate for Geosciences, AISES awarded mini-grants to college chapters to bring the geosciences to children in nearby schools. The AISES college students were excellent ambassadors for science and also enabled the children to see college students who are Native and are science majors. As an added benefit, the geoscience majors who participated in the outreach program gained a stronger sense of commitment to a career in the geosciences, having had the experience of conveying the excitement of their field to the children.

Development of the Leadership Pipeline

As we work toward our goal of increased participation of Native People in the STEM disciplines, we recognize the need to cultivate leaders, within our organization, within our Native communities, and within our disciplines. Leadership development occurs at all stages. In our most successful college chapters, the more senior student members make a point to groom selected underclassmen for future leadership positions in the chapter and in the organization. Some of our professional members, Native scientists and engineers, have taken leadership roles in tribes, serving a chiefs and tribal council members. Native leaders within our profession, such as former ACS President Dr. Nancy Jackson, help to raise the visibility of Native People within the STEM communities.

Cultural Identification

The integration of science and traditional Native culture is a recurring theme throughout AISES activities. Therefore cultural activities are always a part of AISES events. Each National Conference features an Honors Banquet, attended by all, to which many AISES members wear traditional regalia. Honor songs, sung by traditional drum groups, are offered in recognition of award winners and scholarship recipients. Figure 7 depicts a powwow held at the AISES National Conference.

Figure 7. Powwow at the AISES National Conference. (Courtesy of AISES.)

The members of the AISES Council of Elders play key roles within AISES (http://www.aises.org/about/elders_council). At the National Conference, plenary sessions and meals begin with a prayer by a Council of Elders member in his or her Native language. Throughout the conference, our AISES Elders offer blessings, stories, wisdom, and encouragement. Our Elders remind us of traditional Native values and ways and help us to appreciate the connections between Native tradition and scientific pursuits. Photographs of some of our Elders are shown in Figure 8.

Figure 8. AISES Council of Elders members – Left: Andrea and Horace Axtell (Nez Perce); Right: Stan and Cecilia Lucero (Laguna and Acoma Pueblos). (Courtesy of AISES.)

Challenges

In spite of many successes, challenges and barriers remain in our pursuit of greater Native presence in the STEM disciplines.

A major challenge is the dearth of Native faculty members in the STEM fields at colleges and universities across the country. Even on tribal college faculties, there are very few Natives with Ph.D. degrees in the STEM fields. While there has been an increase in the number of Ph.D. degrees awarded to Native Americans in the science and engineering fields, having risen from 78 in 2001 to 103 in 2010, this trend does not translate to an increase in Native Americans in college and university faculty positions in STEM (*1*). This absence of role models does not encourage Native students to declare STEM majors nor to see themselves as future faculty members.

While the efforts of AISES and the American Chemical Society have helped to raise awareness, the lack of visibility of Chemistry in Indian Country, and the lack of visibility of Native Americans in the Chemistry community, persist.

Within the Chemistry discipline, there is also a lack of awareness of, and sensitivity to, Native American cultural issues. The Open Chemistry Collaborative in Diversity Equity (OXIDE) is working with Chemistry Departments across the country to raise awareness of diverse cultures (*2*).

Future Directions

Successful programs to support Native college students in pursuit of academic careers in the STEM fields are necessary for the future advancement of the AISES mission. This is a current priority.

The nationwide AISES network can be leveraged for future outreach efforts. For instance, the K-12 outreach model that was successful in the geosciences could be replicated for Chemistry.

We must continue to seek and cultivate the huge store of untapped talent for science in Indian Country. Scientific knowledge and tools, in Indian hands, with Indian intelligence, creativity, resourcefulness, and values, is a powerful force for good in Indian Country and for the world.

Acknowledgments

The author thanks the staff and Board of Directors of the American Indian Science and Engineering Society for their gracious help with some of the details and also for permission to use AISES images. I give special thanks to AISES Board members Dr. Melinda McClanahan (Choctaw) and Dr. Twyla Baker-Demaray (Three Affiliated Tribes), to AISES CEO Sarah Echohawk (Pawnee), and to AISES staff members Lisa Paz (Pawnee and Comanche) and Shirley LaCourse (Oglala Lakota, Oneida, Yakama, and Umatilla). I extend my gratitude to 2013 ACS President Dr. Marinda Wu for her leadership in spearheading the diversity symposium and for the opportunity to build another little bridge between the two national organizations dearest to my heart, AISES and ACS.

References

1. Nelson, D. J.; Brammer, C. N. *A National Analysis of Minorities in Science and Engineering Faculties at Research Universities*; University of Oklahoma: Norman, OK, 2010.
2. Hernandez, R. *Open Chemistry Collaborative in Diversity Equity*; 2010. Available from: http://www.oxide.gatech.edu/ver1.0/index.html (accessed May 18, 2014).

Chapter 24

Career Opportunity Challenges for Asian American Chemistry Professionals and Scientists

Guang Cao*

Chinese American Chemical Society, P.O. Box 786, Edison, New Jersey 08818
*E-mail: Guang.h.cao@gmail.com

Asian American chemists and chemical professionals face a unique career development challenge. While the most salient issue for most of the other minority groups as a whole is under-representation in the STEM fields, strong evidence indicates that Asian Americans are under-represented in positions of decision-making and influence. In addition to highlighting this issue with data, this article also attempts to explore the root causes and provides suggestions for potential solutions.

Asian Americans in the U.S.

The 2010 US census concludes that the Asian American population increased the fastest among all major race groups during 2000 – 2010, growing by 44% from 10.2 million in 2000 to 14.7 million in 2010 (*1*). (By comparison, Hispanic population grew by 43% in the same time period.) Asian Americans gained the most in share of the total population, moving up from about 4 percent in 2000 to about 5 percent in 2010. Accompanying this trend is the increase in their share of the US workforce from just under 4% in 2000 to slightly over 5% in 2009; see Figure 1.

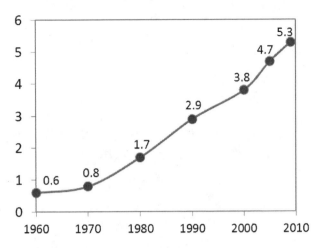

Figure 1. Asian American and Pacific Islanders as Share of US Workforce, 1960-2009. Data source: "Diversity and Change - Asian American and Pacific Islander Workers," Center for Economic and Policy Research, 2011

As is true with all race groups, Asian Americans consist of a diverse group of people with widely varied cultural, educational, and socioeconomic backgrounds. The challenges they face are, therefore, diverse, ranging from poverty, cultural assimilation, and access to quality education on one hand, to career advancement and social influence on the other. In spite of the diversity, perception by the US public of the Asian Americans has been rather uniform. They are portrayed in the media as a "model minority group" for having attained the highest level of education on average among all ethnic groups in the US, see Figure 2.

Asian Americans in the STEM Fields

The professional ranks in the STEM fields have been disproportionately represented by Asian Americans. According to the 2013 National Science Foundation study "Women, Minorities, and Persons with Disabilities in Science and Engineering, Arlington, VA, NSF 13-304, February 2013," Asian Americans represented 18% of the Science and Engineering workforce in 2010 (2).

Membership on the American Chemical Society (ACS) reflects a similar trend: Asian Americans and Pacific Islanders represented almost 15% of the ACS membership in 2007, increasing steadily from 13.4% over a seven year period when such data were examined (3).

Perhaps because of this, Asian Americans, especially Asian American men, are not treated as minorities in the STEM fields. Some US universities in recent years have instituted quota to limit the number of Asian American (men) faculty members, and some elite US universities have been using higher academic admission standards to limit the number of Asian American students on their campuses, a controversial practice that is viewed by some as being discriminatory against Asian American students.

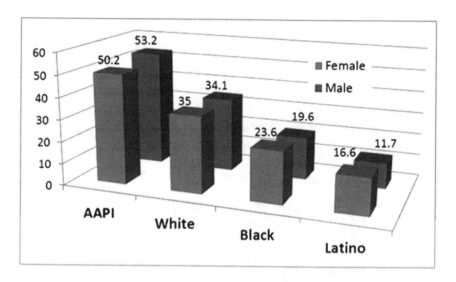

Figure 2. Percent Workers with a College Degree or More, by Ethnicity, Ages 16+. Data source: Center for Economic and Policy Research: analysis of American Community Survey (ACS), 2009. AAPI is abbreviation for Asian Americans and Pacific Islanders

These educational issues aside, there is a bigger issue that has hampered career development of many Asian American chemists and chemical professionals (or Asian American scientists and engineers as a whole) – that is, under-representation of Asian Americans in positions of decision-making and influence.

Under-Representation of Asian American Chemistry Professionals in Positions of Decision-Making and Influence

Although this issue has not received much attention, it is not a new revelation. For example, a 2004 article "American Asians: Achievements Mask Challenges" by Roli Varma (4), pointed out *"the existence of the "glass ceiling" to upward career mobility experienced by Asian Americans in professional occupations."* It further shows that *"despite their good record of achievement, Asian Americans do not reach a level at which they can participate in policy and decision-making responsibilities..."*

Relevant statistics on this gap have been available but consistent and up-to-date reporting has not been the norm, perhaps reflecting the lack of attention to this issue. One piece of data is found in the 2007 National Science Foundation report "Women, Minorities, and Persons with Disabilities in Science and Engineering" (2), which provided Figure 3. As shown in Figure 3, Asian scientists and engineers, together with those of Native Hawaiian and Pacific Islanders, have the lowest median subordinates amongst the minority groups. Unfortunately this type of data was not included in the subsequent 2009, 2011, and 2013 reports (2).

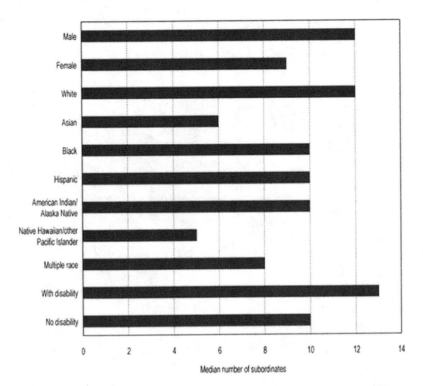

Figure 3. Median number of subordinates of science and engineers employed in business or industry, by sex, race/ethnicity, and disability, and disability status. Source: Woman, Minority and Persons With Disabilities in Science and Engineering, NSF-07-315, page 22 (December 2006), with permission.

Another piece of information is found in ACS annual salary surveys. The author obtained these unpublished data with gracious help from ACS staff (3). As shown in Figure 4 below. Of all ACS members answering the ACS survey, those identified themselves as Asians reported the lowest percentage of being in managerial positions.

Root Causes

It would be far too simplistic and even irresponsible to attribute to bias or discrimination as a major cause for the issue at hand. As any issues concerning race and ethnicity, there are a plethora of factors playing a role. First of all, there is the factor of cultural differences, most prominently in the acceptance and valuation of self-effacement vs. self-promotion. In the US, outspokenness and even braggadocio are not viewed nearly as negatively as in most Asian cultures. The popular American saying "A squeaky wheel gets the grease" contrasts dramatically with the Japanese saying "A nail that sticks out gets hammered" or the Chinese saying "A beam that sticks out gets rotted first." As eloquently pointed out by Prof. Ronald Hoy twenty years ago (5):

"The cultural injunctions that permeate some Asian societies - those values that discourage assertiveness, outspokenness, and competitiveness in groups - work against Asian Americans, especially in the culture of science. Thus, an Asian job candidate may be viewed, at best, by an Anglo-European department chair as being shy or indifferent or, at worst, as having nothing to say or being unable to act decisively."

Another factor is the degree of community involvement and volunteerism, which tends to reflect social integration. According to the latest US Department of Labor statistics (6), among the major race and ethnicity groups, Asian Americans continued to volunteer at a lower rate (19.6 percent) than did white (27.8 percent) and blacks (21.1 percent), but higher than Hispanics (15.2 percent).

On the side of public perception, Asian Americans are the perpetual foreigners, unlike European or Hispanic Americans who are identified as "American" much more readily. A 2008 paper on British Journal of Social Psychology concluded that the general American public viewed the British actress Kate Winslet more American than the America-born actress Lucy Liu (ref). This conclusion may illicit a chuckle at a dinner party but it is no laughing matter if the perception persists. Additionally there are far fewer Asian American stars in American pop culture than blacks and Hispanics.

Finally as Asia has become more competitive economically, there appears to be some ambiguity in public trust of Asian Americans' national affinity and affiliation. According to a 2009 survey by the organization C-100 ("Still The "Other?" – Public Attitudes Toward Chinese and Asian Americans" (7), only 55% and 57% US general public thought that Chinese Americans would support US in an economic and military conflict with China, respectively, whereas 75% and 77% Chinese Americans said that they would support the US. This perception gap highlights the need for raising awareness of this issue to both the general public and the Chinese/Asian Americans.

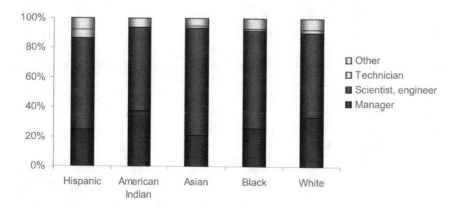

Figure 4. Asian Americans answering the 2007 American Chemical Society annual salary survey repoted the lowest percentage of being in managerial positions. (data provided by Jeffrey Allum of ACS)

As mentioned in the beginning section of this article, the diversity within the Asian American community makes exceptions to the above root cause analysis, which is mostly suitable for those with Confucian cultural backgrounds. Indian Americans, for example, have arguably fared better even though they are a newer group of immigrants than, say Chinese or Korean Americans.

What Can Be Done?

Therefore Asian American scientists / chemists face a unique challenge. On one hand they are "over-represented" in the STEM fields and are sometimes not considered minorities, on the other they are under-represented in positions of decision-making and influence. How should Asian American scientists / chemists, and the society as a whole, address such a challenge? The current social discourse on diversity is dominated by issues related to the under-represented minorities (women, blacks, Hispanics, disabled) and the LGBT community. These are undoubtedly just causes, but is there any room to address the Asian Americans' dilemma on the Diversity and Inclusion platform? Is the issue a transitory one that can take care of itself in due course?

The cause for action is two-folds: one is that this is ultimately an equal-rights issue concerning the American ideal of everyone reaching his/her ultimate potential if he/she endeavors; the other is the worrisome observation that the situation has not remedied itself over the past decades. The situation commented by Prof. Ronald Hoy twenty years ago, for example, still painfully exists in the workplace today.

A multi-prong approach is appropriate for the complex issue at hand. First, there should be more awareness of the issue by the general public. As Wesley Yang wrote in a New York magazine article in 2011, the Asian Americans are not model citizens, they are "paper tigers" who have excelled in education but once the "paper" is done, they fall behind. Along with awareness there should be understanding and appreciation for different cultures and behaviors. Secondly, Asian American scientists/chemists should be the agents for change themselves, be more vocal and assertive in expressing their views and concerns. Along this line it would help a great deal to be more involved in the communities they live and work in, become volunteers. For this purpose the Chinese American Chemical Society has been a great venue for Chinese American chemists. Thirdly, for an organization such as the ACS, there should be a committee or sub-committee dedicated to Asian American chemical professionals. In the ACS governance body, there are committees for women chemists, young chemists, minority chemists (blacks and Hispanics), and senior chemists, but there is not one for the 15% Asian American members.

Concluding Remarks

The issue of career opportunity has confronted Asian American scientists and engineers for some time, yet it has hardly received much attention. It is therefore time to start address the issue in a holistic way, from within Asian American communities to the general public as a whole. The common goal is to reach the American ideal of providing equal opportunities for the realization of everyone's potential, thus maximizing individual contributions to society. In this endeavor, everyone has a role and everyone can make a difference – so get involved!

References

1. 2010 US Census News Release: "2010 Census shows America's Diversity, March 24, 2011, http://www.census.gov/newsroom/releases/archives/2010_census/cb11-cn125.html.
2. Reports can be found at http://www.nsf.gov/statistics/women/.
3. Allum, J. Private communication: unpublished data from "ACS Annual Salary Survey".
4. Varma, R. Asian Americans: Achievements Mask Challenges. *Asian J. Soc. Sci.* **2004**, *32* (2), 290–307.
5. Hoy, R. A Model Minority Speaks Out on Cultural Shyness. *Science* **1993**, 1117–1118.
6. US Department of Labor Economic News Release: "Volunteering in the United States, 2012". http://www.bls.gov/news.release/volun.nr0.htm.
7. 2012 and precious reports by Committee of 100: http://www.survey.committee100.org/.

Chapter 25

Diversity and Inclusion from the Global Perspective

Teri Quinn Gray*

DuPont Crop Protection, P.O. Box 30, Newark, Delaware 19714
*E-mail: Teri-Quinn.Gray@dupont.com

It is well known that in a corporate setting diversity and inclusion have a positive impact on workforce development, employee morale and productivity, and financial performance. Likewise, in the scientific area, it is desirable to recruit, retain and reward the best talent, irrespective of gender, race, ethnicity, national origin, sexual preference, and any other characteristics. As chemical companies are becoming increasingly globalized, it is important also to incorporate the global perspectives to the mix. In fact, a corporate culture and company policies that favor diversity, inclusion, and global awareness are essential for the company's competitiveness and long-term well being. The policies may pertain to hiring, promotion, work assignments, social interactions, and recognition. The author will review the business case for diversity, inclusivity, and global awareness, using DuPont Company and the American Chemical Society (ACS) as examples wherein these practices have been implemented.

Introduction

The U.S. has always had a lot of diversity among its citizens. For example, according to the U.S. Census, in 2010 the U.S. population consists of 16.4% Hispanic or Latino Americans, 12.6% African Americans, 4.6% Asian Americans, and 0.9% Native Americans. The male/female ratio is 48.8 : 51.2. In terms of ancestries, the largest groups (in decreasing order) are German, Irish, African American, English, American, Mexican, Italian, Polish, French, American Indian, and others. We can break down the population further by using additional

© 2014 American Chemical Society

characteristics, but regardless of our differences, we all contribute in various ways to the well-being and the welfare of the nation.

In recent years, both public and private organizations have realized that a diverse workforce is beneficial to the organization's efficiency and productivity (1–3). Many companies have found out that their ability to successfully innovate starts with a diverse workforce complete with high-caliber, high-performing people of various backgrounds, experiences and perspectives. As a result, many of them have instituted diversity and inclusion programs. In addition, the world is increasingly globalized. It has become easier for people, money, information, goods and services to move across national boundaries. It is natural then to extend the diversity concept to the international arena.

The goal of this chapter is to examine diversity, inclusion, and global perspectives as opportunities for growth and excellence for U.S. organizations. Particular examples will be given of DuPont and the ACS, where the author has some familiarity with their diversity and inclusion programs.

Diversity, Inclusion, and Global Perspectives

Because these terms occur frequently, it may be useful to define them in terms of the corporate context.

Diversity

Diversity can be defined as the presence of a variety of experiences and perspectives, including race, culture, religion, mental or physical abilities, heritage, age, gender, sexual orientation, gender identity and other characteristics. Diversity can also be thought of as "other-ness" or those human qualities that are different from our own and outside the groups, to which we belong, yet present in other individuals and groups.

In a corporate setting, a diversity program typically consists of a set of conscious practices:

- Understand and appreciate interdependence of humanity, cultures, and the natural environment;
- Practice mutual respect for qualities and experiences that are different from our own;
- Understand that diversity includes not only ways of being but also ways of knowing;
- Recognize that personal, cultural and institutionalized unconscious bias creates and sustains preferences that may exclude others;
  Build alliances across differences so that we can work together to achieve organizational objectives.

Inclusion

Inclusion means actively promoting community among people and increasing their comfort with diversity. A program involving inclusion may entail the following elements:

- Recognize, value, and affirm all aspects of diversity;
- Seek active, intentional, and ongoing engagement with people and in their communities;
- everyone feels respected and appreciated – so we can individually and collectively do our best work;
- Replace the old adage "Treat others as you want to be treated" with "Treat others as they want to be treated."

Global Perspectives

As the world is increasingly globalized, there are more interactions among people from diverse cultures, beliefs, and backgrounds than ever before. A global perspective is necessary in this changing world order in which economic, social, political, cultural and environmental factors are part of a global community and not restricted in national boundaries.

The Case for Diversity, Inclusion, and Global Awareness

First of all, it may be noted that diversity is a business reality; it is an unavoidable component of the 21st-century world. If we look at the Census Bureau projections, white non-Hispanic Americans will constitute less than 50% of the U.S. population by 2050. Thus, the workforce will be increasingly diverse by mid-century. Likewise, the racial and ethnic mix of consumers will also change. As globalization continues, global manufacturing, markets, and manpower will play more and more significant roles. The trends today is towards greater connectivity and interdependence, among individuals, among ethnic groups, and among countries.

It has been known for a long time that positive co-worker relationships and respect for the individual dignity are important factors in fostering a positive workplace environment. Thus, a corporate culture that emphasizes teamwork, mutual respect, and individual dignity tends to promote better worker morale and enhances worker efficiency and productivity. Diversity and inclusion are part of the same factors that contribute to improved worker engagement. Because the demographics of consumers will change in the future, if a company is diversity-savvy, it will be more capable of building intimacy between the company and its customers, and between its employees and its customers (*1–3*).

In the chemical industry in particular, competition is very keen. A more diverse workforce and the ability to attract the best talent in the world is a distinct competitive advantage. Engaged and creative employees can facilitate innovation and accelerate speed to market. An inclusive and global corporate culture can attract and retain best available talent from "global" pools. This is true in industry as well as in academia (4).

If we look at a bigger picture and ask about the world's greatest challenges today, we are likely to hear about global water crisis, climate change, and population explosion. Each of these challenges has a strong international component. If a company wants to develop products or services related to these challenges, global awareness and cultural competence would be important success factors.

Why not Yet?

If the case for diversity, inclusion, and global awareness is so strong, why is the diversity program not broadly accepted and implemented in corporate America?

Let us look at the metaphor of the iceberg (Figure 1). Because 10% of the iceberg is above the water line, we can call it "Visible Diversity." Too often that 10%, what we see/notice first, determines how we behave. And this can have many consequences. None of us would make a loan decision based on having only 10% of the applicant's financial data, yet we make these interpersonal decisions having only the tip of the iceberg – visual data. What happens if we go beneath the waterline and explore the other 90% (just a partial list being shown in Figure 1). We may discover things that we did not consider before or even knew existed. These are the "connections" that can make all the difference in how we interact with someone. Many of our feelings and biases are implicit, such that we are not consciously aware of them.

Thus, human prejudices are widespread and often hidden among all of us. It takes a lot of effort to bypass these feelings and biases. Although a company may have the best of intentions in terms of diversity, it cannot change human nature quickly. Thus, the success of a diversity program depends on the individuals as much as the group. It may not be easy to know what different groups or individuals really need. To be effective, a diversity program needs to be customized for a given organization, or even a group within an organization.

Note that the benefits of diversity are not automatic and do not simply occur from a diverse body. An organization must be intentional, persistent, and strategic. Furthermore, a program like diversity depends on collaboration among many people. Sometimes collaboration is harder than compromise. In general, tolerance is the first step toward mutual respect. Good leaders are needed to provide guidance, encouragements, mentorship, and courage.

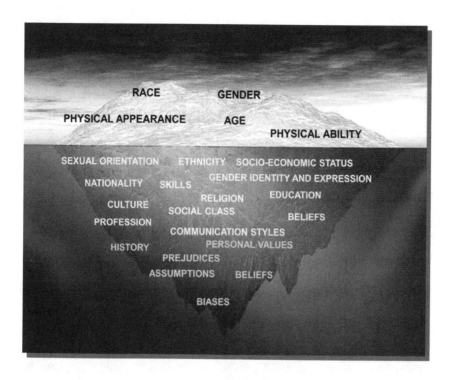

Figure 1. An iceberg as a metaphor for human prejudices (5). (Courtesy of the American Chemical Society.)

Example of DuPont

At DuPont, diversity is viewed as an intergral part of its continued success. Over the years, the company has hired many talented employees from all over the world, irrespective of gender, race, ethnic origin, and nationality. On their website, the company states that "Respect for people and our employees has been one of DuPont's core values since its inception. DuPont diversity brings individuals of all backgrounds together to pool their talent and creativity. Here we believe that global excellence is best achieved by reflecting a global diversity in our employees" (6).

Because DuPont is a technology-driven company, it is understood that the company can only be successful with the full commitment, participation, creativity, energy and cooperative spirit of its diverse workforce. It is DuPont's policy not to discriminate against any employee or applicant for employment because of age (within statutory limits) race, religion, color, sex, disability, national origin, ancestry, marital status, sexual orientation, gender identity/expression or veteran status. In addition, there are many diversity networks for DuPont employees within the company.

Because of DuPont's track records, it has garnered quite a few awards for their diversity policies and practices. For example, DuPont Pioneer has been selected as the 2013 recipient of the Greater Des Moines Partnership's Diversity Award for promoting diversity at their businesses and organizations. The National Association for Female Executives (NAFE) named DuPont as a leader in the U.S. in their commitment to female leadership. As yet another example, DuPont (recently retired)Vice President and Assistant General Counsel Hinton J. Lucas, Jr. recently was honored by several prominent organizations for his leadership in promoting and fostering diversity in the legal profession.

At DuPont Crop Protection where the author works, crop protection tools are designed to help growers feed the world. DuPont believes that by working together, they, along with their customers and partners, can find better ways to improve the quantity, quality and sustainability of the world's food supply. DuPont combines global innovation with local collaboration to turn ideas into opportunities. A diverse corporate culture and strong global perspectives are highly valued assets.

Example of ACS

The American Chemical Society is a 501(c)3 non-profit organization chartered by the U.S. Congress in 1937 to advance chemistry in all its branches, promote scientific research and inquiry, and foster public welfare and education. From my personal experience as a volunteer, ACS seems to have a strong culture in diversity and inclusion (*7, 8*). It is not perfect, and improvements can still be made, but it seems to be working. Thus, for many years ACS has the following committees to promote diversity and inclusion within the Society: Women Chemists Committee (WCC), Younger Chemists Committee (YCC), and Committee on Minority Affairs (CMA), and Committee on Chemists with Disabilities CWD). Moreover, an umbrella group for all diversity and inclusion groups was started in 2006 as Collaborative Working Group, which became Joint Subcommittee on Diversity in 2007, and is now called ACS Diversity & Inclusion Advisory Board (D&I). The current constituent members of D&I include WCC, YCC, CMA, CWD, SCC, as well as Senior Chemists Committee (SCC), Committee on Professional Training, (CPT), Committee on Technician Affairs (CTA), Division on Professional Relations (PROF), and Diversity Partners. The vision, mission, and goals of D&I are given in Figure 2.

The various groups within D&I work hard to leverage each others' work and outreach beyond ACS to strengthen efforts to bolster a more inclusive environment within the ACS and broader chemical enterprise. D&I organizes and conduct a wide variety of events and activities at the two ACS national meetings held each year. In every meeting, there is at least one symposium on diversity and inclusion. In fact, this chapter (and part of this book) result from one such symposium, organized by the Committee on Minority Affairs at the ACS national meeting in Indianapolis in September 2013.

> **VISION**
> The American Chemical Society is an inclusive community of highly skilled chemical professionals that reflects the diversity of the workforce today.
>
> **MISSION**
> The mission is to promote and advance diversity within and on behalf of the Society.
>
> **GOALS**
> - Increase the diversity of ACS membership and governance with emphasis on African Americans, Hispanics, Native Americans and younger chemists
> - Advance diversity and inclusion in the chemical sciences by raising awareness among ACS members, students and relevant stakeholders
> - Recognize and disseminate information on outstanding achievements in diversity by members, staff and other stakeholders

Figure 2. Vision, Mission, and Goals of ACS Diversity and Inclusion Advisory Board

In terms of awards, D&I provide the Award in Recognizing Underrepresented Minorities in Chemistry for Excellence in Research & Development. The purpose of the award is to recognize and promote the participation of under-represented minorities in chemistry by identifying and recognizing a distinguished minority chemist who has made significant contributions to chemical research. ACS provides the Stanley C. Israel Regional Award that recognizes individuals and/or institutions who have advanced diversity in the chemical sciences and significantly stimulated or fostered activities that promote inclusiveness within the region.

In addition, WCC provides several awards specifically for women, including WCC/Eli Lilly Travel Award, WCC Overcoming Challenges Award, and WCC Rising Star Award. YCC provides the YCC/CIBA Young Scientist Travel Award and YCC Leadership Development Award.

Finally, ACS has issued a public statement on diversity as given in Figure 3. The statement clearly highlights diversity and inclusion as an ACS core value and points out some of the advantages of diversity and inclusion, including global competitiveness, innovation, manpower and community.

More information on the ACS diversity and inclusion programs can be found in two articles at *Chem. Eng. News* (*9, 10*) and in a presentation given by the author at the ACS national meeting in Indianapolis in 2013 (*11*).

> The American Chemical Society believes that to remain the premier chemical organization that promotes innovation and advances the chemical sciences requires the empowerment of a diverse and inclusive community of highly skilled chemical professionals regardless of race, gender, age, religion, ethnicity, nationality, sexual orientation, gender expression, gender identity, presence of disabilities, educational background, and other factors. Chemical scientists rely on the American Chemical Society to promote inclusion and diversity in the discipline.
>
> To enable scientific progress and maintain its global competitive edge, the American Chemical Society remains committed to inspiring and educating the present and future generations of diverse, innovative, and creative chemical professionals. By promoting inclusion and equity to all, the American Chemical Society will succeed in fostering a diverse community of professionals in the chemical sciences who will be the catalyst for transforming the world through their full participation and integration into the chemical professions.

Figure 3. ACS Statement on Diversity. Approved by the ACS Board of Directors - 2007

Conclusions

In this chapter, the author has summarized the business case for diversity, inclusivity, and global awareness, using DuPont Company and the American Chemical Society (ACS) as examples. As noted earlier, for a diversity program to succeed, an organization must be intentional, persistent, and strategic. Good leadership is needed to provide guidance, encouragements, mentorship, and courage.

The Diversity Creed, written by Gene Griessman in 1993 (*12*), captures the spirit of diversity and provides sage advice to all of us in our dealings with one another:

> I believe that diversity is a part of the natural order of things—as natural as the trillion shapes and shades of the flowers of spring or the leaves of autumn. I believe that diversity brings new solutions to an ever-changing environment, and that sameness is not only uninteresting but limiting.
>
> To deny diversity is to deny life—with all its richness and manifold opportunities. Thus, I affirm my citizenship in a world of diversity, and with it the responsibility to....
>
> • Be tolerant. Live and let live. Understand that those who cause no harm should not be feared, ridiculed, or harmed—even if they are different.
>
> • Look for the best in others.

- Be just in my dealings with poor and rich, weak and strong, and whenever possible to defend the young, the old, the frail, the defenseless.

- Avoid needless conflicts and diversions, but be always willing to change for the better that which can be changed.

- Seek knowledge in order to know what can be changed, as well as what cannot be changed.

- Forge alliances with others who love liberty and justice.

- Be kind, remembering how fragile the human spirit is.

- Live the examined life, subjecting my motives and actions to the scrutiny of mind and heart so to rise above prejudice and hatred.

- Care. Be generous in thought, word, and purse.

References

1. Anon. Global Diversity and Inclusion: Fostering Innovation Through a Diverse Workforce. *Forbes Insight*, 2011. http://images.forbes.com/forbesinsights/StudyPDFs/Innovation_Through_Diversity.pdf.
2. Robinson, M.; Pfeffer, C.; Buccigrossi, J. *Business Case for Inclusion and Engagement*; wetWare, Inc.: Rochester, NY, 2003. http://workforcediversitynetwork.com/docs/business_case_3.pdf.
3. Anon. What is the business case for diversity? http://www.workforce.com/articles/20086-whats-the-business-case-for-diversity.
4. Anon. Academic Diversity: A Look at Race, Ethnicity, and Gender in Higher Education. http://www.acs.org/content/acs/en/policy/acsonthehill/briefings/academic-diversity.html.
5. Lolita, C. The Chandler Group. Figure 1 is taken from the presentation, "ACS Diversity Dialogue Session:Setting the Stage, Starting the Dialogue."
6. Anon. DuPont Diversity – The Key to Our Success. http://www.dupont.com/corporate-functions/careers/why-dupont/articles/diversity.html.
7. Wang, L. Diversifying Chemistry Faculty. *Chem. Eng. News* **2011**, *89* (9), 46–47.
8. Lane, T. H.; Francisco, J. S. Building A Diverse Profession And Inclusive Community. *Chem. Eng. News* **2010**, *88* (25), 35.
9. Schmidt, D. G.; Gray, T. Q. Advancing A More Diverse Profession. *Chem. Eng. News* **2011**, *89* (6), 36.
10. Moore, M. K.; Coffman, M. B. Committee on Technician Affairs: 50 Years and Counting. *Chem. Eng. News* **2013**, *91* (50), 30.
11. Gray, T. Q. Diversity and Inclusion from the Global Perspective. Presented at the 246th ACS National Meeting, Indianapolis, IN, Aug 2013.

Chapter 23

American Indian Science and Engineering Society (AISES): Building a Successful Model for Diversity and Inclusion

Mary Jo Ondrechen*

Department of Chemistry and Chemical Biology, Northeastern University, 360 Huntington Avenue, Boston, Massachusetts 02115
*E-mail: m.ondrechen@neu.edu

In this chapter, the work of the American Indian Science and Engineering Society (AISES) is described. AISES, a national organization of Native American scientists and engineers and headquartered in Albuquerque, NM, is 3000 members strong. The strength of the science talent pool among Native Americans, Alaska Natives, First Nations, and Native Hawaiians is discussed. Strategies for overcoming cultural barriers and for success in diversity and inclusion are also included. Moreover, examples of successful synergy between Chemistry and Native America are presented.

About AISES and Its Mission

The American Indian Science and Engineering Society (AISES) was established in 1977 with the mission to "substantially increase the numbers of American Indians and Alaska Natives" in the science and engineering fields. AISES is headquartered in Albuquerque, New Mexico and its membership extends across the United States and Canada.

AISES is a non-profit, professional society. The voting members of AISES are Native professionals with degrees in a Science, Technology, Engineering, or Mathematics (STEM) discipline. For AISES membership purposes, "Natives" are defined as all of the Indigenous People of North America. Thus American Indians, Alaska Natives, Native Hawaiians, and First Nations (Indigenous People of Canadian origin) are all part of the AISES professional membership and in this chapter are referred to simply as Natives or Indigenous People. Persons who do

© 2014 American Chemical Society

12. Griessman, B. E. *Diversity: Challenges and Opportunities*; Harper Collins: New York, 1993.

Editors' Biographies

H. N. Cheng

H. N. Cheng (Ph.D., University of Illinois) is currently a research chemist at Southern Regional Research Center of the U.S. Department of Agriculture in New Orleans, where he works on projects involving improved utilization of commodity agricultural materials, green chemistry, and polymer reactions. Prior to 2009 he was with Hercules Incorporated where he was involved (at various times) with new product development, team and project leadership, new business evaluation, pioneering research, and supervision of analytical research. Over the years, his research interests have included NMR spectroscopy, polymer characterization, biocatalysis and enzymatic reactions, functional foods, pulp and paper technology, and green polymer chemistry. He is an ACS Fellow and a POLY Fellow and has authored or co-authored 186 papers, 24 patent publications, co-edited 10 books, and organized or co-organized 24 symposia at national ACS meetings since 2003.

Sadiq Shah

Sadiq Shah (Ph.D., Washington University, St. Louis) is the Vice Provost for Research at the University of Texas – Pan American with responsibilities for managing, directing, and growing the research, scholarship, and creative activities as well as technology transfer efforts on campus. Earlier Dr. Shah served successively as Associate Vice President for Research at California State University Channel Islands, Associate Vice President for Research and Economic Development at Western Kentucky University, Director of the Western Illinois Entrepreneurship Center and the Office of Technology Transfer at Western Illinois University, Manager, Product & Technology Development for Health Care markets at STERIS, BMS and Merck, and senior research chemist at Petrolite Corporation. He is an ACS Fellow and has been responsible for guiding the development of 20 new products from concept to launch and 7 technology platforms. He has 15 patents, has edited and co-edited two books, written book chapter, and published 30 research articles and other articles related to technology transfer.

Marinda Li Wu

Marinda Li Wu (Ph.D., University of Illinois) is the 2013 President of the American Chemical Society (ACS). She has over thirty years of industrial experience at Dow Chemical R&D and Dow Plastics Marketing and additional entrepreneurial experience with various small chemical companies and startups

© 2014 American Chemical Society

including "Science is Fun!" which she founded to engage young students in science and enhance public awareness. She has served in many leadership roles at local and national ACS levels. She was elected to the ACS Board of Directors in 2006 and served as Director-at-Large until 2011. In 2011, she was elected to the ACS Presidential succession, where she was invited to give plenary lectures worldwide and was made an honorary member of both the Romanian Chemical Society and the Polish Chemical Society. She also serves on the University of Illinois Chemistry Alumni Advisory Board, the International Advisory Board for the 45th IUPAC World Chemistry Congress 2015, and the Board of Directors for the Chinese-American Chemical Society. She holds 7 U.S. patents and has published a polymer textbook chapter and numerous articles in a variety of journals and magazines.

Indexes

Author Index

Baum, R., 129
Beattie, T., 135
Brewer, J., 183
Brown, R., 143
Bryant-Friedrich, A., 245
Cadwalader, E., 245
Cao, G., 265
Chao, J., 93
Cheng, H., xiii, 1
Confalone, P., 163
Fraser, E., 143
Fraser, J., 107
Frishberg, M., 69
Gray, T., 273
Harwell, D., 143
Hernandez, R., 207
Herring, C., 225
Kruse, A., 183

Laurence, J., 39
Lawrence, N., 183
Lechleiter, J., 31
Omberg, K., 61
Ondrechen, M., 255
Ornstein, J., 173
Polk, K., 143
Rich, R., 1, 115
Rodriguez, G., 83
Shah, S., xiii, 1, 149
Shulman, J., 49
Tilstra, L., 237
Vercellotti, J., 195
Vercellotti, S., 195
Watt, S., 207
Wu, M., xiii, 1
Zhou, H., 173

Subject Index

A

Alternative chemistry careers
  career change, leaving laboratory, 98
  career decision points, 94
    for author, 97f
  division of competitive industry sectors for IBM, 100f
  innovation using intellectual property assets, 104f
  instrumentation, 96
  intellectual property licensing at IBM approaches, 98f
    Bad Cop approach, 99f
    Good Cop approach, 100f
  introduction, 93
  licensing of software tool, 102f
  manufacturing gene chips, 102f
  non-semiconductor industries, manufacturing, 103f
  personal background as chemist, 94
  prospective out-of-industry commercialization ventures, 101
  summary, 105f
  using intellectual property assets, commercial innovation, 104
American Chemical Society (ACS) entrepreneurial initiative
  ACS's first-ever Entrepreneur Summit, 146
  conclusions and future directions, 147
  introduction, 143
  overview, 1
  Presidential Task Force, innovation and job creation, 144
  results of pilot, 145
  Task Force recommendations, 144
American Indian Science and Engineering Society (AISES)
  2013 AISES Leadership Summit, 259f
  AISES National Conference, 260
  award-winning AISES college chapter, 258f
  college programs, access and success, 257f
  future directions, 263
  information about and mission, 255
  membership, 256
  mission, purpose, 256
  pre-college programs, awareness and retention, 256
  professional programs, leadership and change, 259
  programs, 256
  projects presented by K-12 students, 258f
Asian American chemistry professionals and scientists
  career opportunity challenges, 265
  2007 American Chemical Society annual salary survey, 269f
  Asian Americans in STEM fields, 266
  Asian Americans in the U.S., 265
  complex issue, multi-prong approach, 270
  conclusions, 270
  positions of decision-making and influence, under-representation, 267
  root causes, 268

C

Capital funds, 175
Career advancement, 107
  against circumstances, pick yourself up and move on, 110
  cut losses and move on, 112
  decision making, 107
  good health and prosperity for family, 113
  keep eyes and ears open always, 114
  know when to hold 'em, to fold 'em and to walk away, 112
  seizing opportunities, 108
  turning sales into a sustainable business, 111
  viewpoint of others, 109
Career opportunities
  academic administrative career, 7
  academic teaching career, 5
  ACS academic members working in different academic institutions, 6t
  breakdown of industrial work functions in chemical industry, 11t
  career in government, 13
  industrial career, 8
  non-traditional careers
    forensic science, 14
    funding agencies, 16
    health, environment, and safety, 14

opportunities involving patents, 15
opportunities involving venture capital, 15
professional organizations, 16
science journalism, 17
science policy and diplomacy, 16
technology transfer, 14
research and development (R&D) laboratories, 10
small businesses and entrepreneurship, 12
type of industrial employers, 9*t*
Chemist strategizing for chemists
associations as profession, 126
career and professional development, 122
conclusions, 127
explore opportunities, 116
extracurricular activities at Berkeley, 117*f*
NIH/Unpaid AAAS internship, year of postdoc, 119*f*
postdoctoral experience, guidelines, 118*f*
postdoctoral fellowship, 116
preparing for career in associations, 115
research grants administration, 122
science policy, 121
science policy work, 118*f*
some other ACS staff with science degrees, 126*f*
strategic plan, 124*f*
strategy development, 122
process, 123*f*
some interesting projects, 125*f*
University of California, Berkeley, 117*f*
working in associations, 120
Chemistry role, 31
chemist participation in biomedical research, 33
conclusion, 37
meeting global challenges, 32
study of chemistry, building on skills developed, 35
Connecting with most powerful ally of science
ACS business connection, six decades, 88
breaking academic Silos, 88
conclusion, 89
education, business and STEM jobs, 87
empowering STEM role in society, 87
helping students become future leaders, 89
introduction, 83
local and global outreach, 89
new world order, adapting, 85

population, 84
rescue of material world, chemistry, 85
rise of prolific and voracious populations, 84
STEM decline requires drastic remake, 86
Creativity
conclusions, 161
definition, 149
dynamics, 151*f*
process
breakthrough ideas, 151
incremental ideas, 150
Creativity, innovation and entrepreneurship nexus, 149

D

Departmental rankings in chemistry, 228
value-in-diversity perspective
proponents, 229
skeptics, 230
Diversity, 274
Diversity and departmental rankings in chemistry, 225
chemistry and chemical engineering programs, characteristics, 233*t*
data and methods, 231
discussion and conclusions, 235
introduction, 226
regression models, 234*t*
results, 232
Diversity and inclusion, global perspective
ACS Diversity and Inclusion Advisory Board, vision, mission, and goals, 279*f*
ACS statement on diversity, 280*f*
case for, 275
conclusions, 280
example of ACS, 278
example of DuPont, 277
introduction, 273
not broadly accepted and implemented in corporate America, reason, 276
Diversity and inclusion in chemistry departments
academic jobs, rare events, 214
broader context, 210
comparative demographics, 213*f*
conclusions, 220
diversity
defined, 209
demographics, 211
diversity equity, other barriers, 215

introduction, 208
meritocracy and bias, 214
Open Chemistry Collaborative in Diversity Equity, 217
　activities, 218
　faculty demographics surveys, 218
　NDEW2013, 218
　recommendations, 219
OXIDE's partnership model, 209*f*
rationale for diversity, 211
removing barriers, 216
top-down approach, 207
Diversity in chemistry, 227

E

End of arc of career
　break from academe, 130
　C&EN's Editor-in-chief, 131
　interviews, 130
　job posting at C&EN, 131
　managing editor, 131
　from medical school dropout to Editor-in-chief of C&EN, 129
　NASW members, 133
　needed job, 130
　publications employing two or more NASW members in 2014, 132*t*
　wanting to be a writer, 130
　West Coast position, 131
Entrepreneurship, 156
　conclusions, 161
　corporations born during past recessions, 157*f*
　global research and development expenditures, 160*f*
　NSF data illustrating R&D performers in different roles, 158*f*
　technology development, U.S. leadership role, 159
　U.S. educational institution, 159
　U.S. federal research and development investment, trends, 160*f*

G

Global chemistry enterprise
　ACS members by categories, 3*t*
　career challenges, 3
　and opportunities, 1
　chemistry employment data, 4*t*
　introduction, 2
　unemployment for all degree levels in chemistry-related jobs, 4*t*
Global perspectives, 275
Government sector, career opportunities
　conclusions, 67
　Congressional Research Service, 63
　Federal Agencies, 63
　Institute of Medicine, 65
　introduction, 61
　National Academies of Science & Engineering, 65
　National Research Council, 65
　salary and benefits, 66
　state and local governments, 66
　STEM professionals, 64
　United States Congress, 62

I

Improving transparency and equity in scholarly recognition
　addressing committee diversity and other best practices, 253
　conclusions, 254
　develop diverse pool of candidates, 251
　evaluate nominees objectively, 252
　evaluation of National Academy of Sciences influence, 253
　increasing awareness of issues within community, 252
　increasing diversity of nomination pool, 253
　inoculation against implicit bias, 248
　introduction, 245
　methods, 247
　percentage of scholarly awards presented to women, 247*t*
　percentage of teaching and service awards presented to women, 248*t*
　review and update awards portfolio, 250
　scholarly awards, 246
　technical disciplinary societies, 246
　women in chemistry community, 246
Inclusion, 275
Industrial careers
　fast start in industry
　　becoming adept at networking, 55
　　communicating effectively, 56
　　learning company culture, 55
　　performance reviews, 59
　　SMART objectives, 58*t*
　　SOPPADA, good template for the written document, 57*f*
　　working hard and smart, 58

obtaining a job and succeeding in industry, 49
resume *vs.* CV, 50
summary, 60
Innovation
  conclusions, 161
  cross-disciplinary technology development, 154*f*
  definition, 152
  evolution of technology, portable audio devices, 153*f*
  growth of well defined ideas with time, 156*f*
  random thoughts evolving into well defined ideas, 155*f*
Innovation and entrepreneurship in the chemical enterprise, 163
  barriers to formation and growth in chemical start-ups, 170*f*
  industrial R&D spending, 169*f*
  job creation *vs.* job loss, 1977-2005, 166*f*
  mega trends and growth opportunities, 168*f*
  Presidential Task Force report, 166*f*
  selected chemical innovations by decade, 165*t*
  statue of chemist holding test tube outside EPCOT, 164*f*
  trends in U.S.-based chemistry jobs, 1989-2009, 167*f*

J

Jobs and career development, ACS resources, 19
  career services
    career consulting services & fairs, 20
    career pathway workshop series, 20
  conclusions, 24
  leadership development, 21
  market intelligence, 22
  professional education, 21
  wealth of other resources, 22

M

Market realities
  innovation, 177
  J-Curve, 177
  sales frameworks, 177

N

National Diversity Equity Workshops (NDEWs), 208
NDEWs. *See* National Diversity Equity Workshops (NDEWs)

O

Open Chemistry Collaborative in Diversity Equity (OXIDE), 208
OXIDE. *See* Open Chemistry Collaborative in Diversity Equity (OXIDE)

R

Rare earth extraction and separation
  academic innovation, 183
  business partnerships, 191
  catalyst use, 187
  extraction, complex problem, 187
  extraction needs quantum jumps, 186
  global demand in different technology sectors, 185*f*
  initial research and funding, 188
  introduction, 183
  lab-scale, pilot-scale and industrial-scale reactions, 190
  method for extracting elements, 187
  opportunities and challenges, 191
  rare earth elements, significance in modern technology, 184
  rare earths and their oxides, 185*f*
  serendipity, 187
  startup company, 189
  supply and demand of rare earth elements, 186*f*
Rose-Hulman Institute of Technology promoting diversity and inclusivity, 237
  curriculum development, 241
  designing workshops, important considerations, 243*t*
  Developmental Model of Intercultural Sensitivity (DMIS), 240
  plans for assessment of campus climate, 243
  response to workshops, 242
  role of Center For Diversity, 239
  the setting, 238
  workshops, content and format, 242

S

Seniors, consulting
  avenues to explore, 136
  business issues, 139
  how to start, 137
  introduction, 135
  some differences, 136
  some examples, 138
  some pros and cons, 137
Successful career in evolving institution
  academic career, success factors
    price of admission, 43
    successful advancement, 44
    sustainability, 45
  academic mission, 40
  academic opportunities and considerations, 39
  career path, 41
  conclusions, 46
  introduction, 39
  scientific discovery, 41
Successful interviewing for industry, 50
  behavioral interviewing, 53
  key performance factors
    capacity, 52
    collaboration, 52
    innovation, 52
    leadership, 52
    risk taking, 52
    solutions, 52
    technical mastery, 51
Successful model for diversity and inclusion, AISES elements
  challenges, 263
  cultural identification, 262
  leadership pipeline development, 262
  youngest members of community, engagement, 261

T

Technical entrepreneurship serving industry
  conclusions, 203
  economic significance of job creation in U.S., 196
  foundations and consulting organizations, 199
  introduction, 195
  model of small chemical business startup, 200
    computer and on-line communication, 202
    laboratory building, design, 201
    modern instrumentation, 201
    promoting our business, 202
    utilities, 201
  small science or engineering businesses, funding sources, 197
  United States Small Business Administration, 198
  venture capital as significant source of funding, 197
  V-LABS, Inc., 200

U

Unexpected career transition
  acknowledge American Chemical Society, temporary tangent, 73
  business development, component, 75
  career beginnings, 70
  defining business development, 74
  final step out of lab, 73
  job description for business development, 81
  lab to career in business development, 69
  learning along way, 76
    communication, 77
    project selection, 77
    travelling, 78
    trip, 79
  pluses and minuses, 80
  project assignment, 72
  setting stage, 71

V

VC market realities, 175
Venture capital funds, 176
Venture capitalist planning, 173
  conclusions, 179
  cost management, 178
  the J Curve, 174
    private investment, 175f
  market realities, 177
Venture capitalists, 176

W

Working in associations
  skills required
    big picture thinking, 120
    creativity, 120

desire to work with other people, 121
mission-driven *vs.* profit-driven, 121
problem solving, 120
written and oral communication, 120
Workplace diversity and inclusion, 17